EMERGING COSMOLOGY

CONVERGENCE

Founded, Planned, and Edited by
RUTH NANDA ANSHEN

EMERGING COSMOLOGY

BERNARD LOVELL

1981
COLUMBIA UNIVERSITY PRESS
NEW YORK

Library of Congress Cataloging in Publication Data

Lovell, Bernard, Sir, 1913–
Emerging Cosmology

(Convergence; v. 1)
Includes bibliographical references and index.
1. Cosmology.   I. Title.   II. Series.
QB981.L873   523.1   81–1860
ISBN 0-231-05304-5   AACR2

Columbia University Press
New York   Guildford, Surrey

Printed on permanent and durable acid-free paper

Contents

Convergence
by
Ruth Nanda Anshen

"There is no use trying," said Alice; "one *can't* believe impossible things."

"I dare say you haven't had much practice," said the Queen. "When I was your age, I always did it for half an hour a day. Why, sometimes I've believed as many as six impossible things before breakfast."

This commitment is an inherent part of human nature and an aspect of our creativity. Each advance of science brings increased comprehension and appreciation of the nature, meaning and wonder of the creative forces that move the cosmos and created man. Such openness and confidence lead to faith in the reality of possibility and eventually to the following truth: "The mystery of the universe is its comprehensibility."

When Einstein uttered that challenging statement, he could have been speaking about our relationship with the universe. The old division of the Earth and the Cosmos into objective processes in space and time and mind in which they are mirrored is no longer a suitable starting point for understanding the universe, science, or ourselves. Science now begins to focus on the convergence of man and nature, on the framework which makes us, as living beings, dependent parts of nature and simultaneously makes nature the object of our thoughts and actions. Scientists can no longer confront the universe as objective observers. Science recognizes the participation of

1

man with the universe. Speaking quantitatively, the universe is largely indifferent to what happens in man. Speaking qualitatively, nothing happens in man that does not have a bearing on the elements which constitute the universe. This gives cosmic significance to the person.

Our hope is to overcome the cultural *hubris* in which we have been living. The scientific method, the technique of analyzing, explaining, and classifying, has demonstrated its inherent limitations. They arise because, by its intervention, science presumes to alter and fashion the object of its investigation. In reality, method and object can no longer be separated. The outworn Cartesian, scientific world view has ceased to be scientific in the most profound sense of the word, for a common bond links us all—man, animal, plant, and galaxy—in the unitary principle of all reality. For the self without the universe is empty.

This universe of which we human beings are particles may be defined as a living, dynamic process of unfolding. It is a breathing universe, its respiration being only one of the many rhythms of its life. It is evolution itself. Although what we observe may seem to be a community of separate, independent units, in actuality these units are made up of subunits, each with a life of its own, and the subunits constitute smaller living entities. At no level in the hierarchy of nature is independence a reality. For that which lives and constitutes matter, whether organic or inorganic, is dependent on discrete entities that, gathered together, form aggregates of new units which interact in support of one another and become an unfolding event, in constant motion, with ever-increasing complexity and intricacy of their organization.

Are there goals in evolution? Or are there only discernible patterns? Certainly there is a law of evolution by which we can explain the emergence of forms capable of activities which are indeed novel. Examples may be said to be the origin of life, the emergence of individual consciousness, and the appearance of language.

The hope of the concerned authors in CONVERGENCE is

that they will show that evolution and development are interchangeable and that the entire system of the interweaving of man, nature, and the universe constitutes a living totality. Man is searching for his legitimate place in this unity, this cosmic scheme of things. The meaning of this cosmic scheme —if indeed we can impose meaning on the mystery and majesty of nature—and the extent to which we can assume responsibility in it as uniquely intelligent beings, are supreme questions for which this Series seeks an answer.

Inevitably, toward the end of a historical period, when thought and custom have petrified into rigidity and when the elaborate machinery of civilization opposes and represses our more noble qualities, life stirs again beneath the hard surface. Nevertheless, this attempt to define the purpose of CONVERGENCE is set forth with profound trepidation. We are living in a period of extreme darkness. There is moral atrophy, destructive radiation within us, as we watch the collapse of values hitherto cherished—but now betrayed. We seem to be face to face with an apocalyptic destiny. The anomie, the chaos, surrounding us produces an almost lethal disintegration of the person, as well as ecological and demographic disaster. Our situation is desperate. And there is no glossing over the deep and unresolved tragedy that fills our lives. Science now begins to question its premises and tells us not only what *is,* but what *ought* to be; *pre*scribing in addition to *de*scribing the realities of life, reconciling order and hierarchy.

My introduction to CONVERGENCE is not to be construed as a prefatory essay to each individual volume. These few pages attempt to set forth the general aim and purpose of this Series. It is my hope that this statement will provide the reader with a new orientation in his thinking, one more specifically defined by these scholars who have been invited to participate in this intellectual, spiritual, and moral endeavor so desperately needed in our time. These scholars recognize the relevance of the nondiscursive experience of life which the discursive, analytical method alone is unable to convey.

The authors invited to CONVERGENCE Series acknowledge a structural kinship between subject and object, between living and nonliving matter, the immanence of the past energizing the present and thus bestowing a promise for the future. This kinship has long been sensed and experienced by mystics. Saint Francis of Assisi described with extraordinary beauty the truth that the more we know about nature, its unity with all life, the more we realize that we are one family, summoned to acknowledge the intimacy of our familial ties with the universe. At one time we were so anthropomorphic as to exclude as inferior such other aspects of our relatives as animals, plants, galaxies, or other species—even inorganic matter. This only exposed our provincialism. Then we believed there were borders beyond which we could not, must not, trespass. These frontiers have never existed. Now we are beginning to recognize, even take pride in, our neighbors in the Cosmos.

Human thought has been formed through centuries of man's consciousness, by perceptions and meanings that relate us to nature. The smallest living entity, be it a molecule or a particle, is at the same time present in the structure of the Earth and all its inhabitants, whether human or manifesting themselves in the multiplicity of other forms of life.

Today we are beginning to open ourselves to this evolved experience of consciousness. We keenly realize that man has intervened in the evolutionary process. The future is contingent, not completely prescribed, except for the immediate necessity to evaluate in order to live a life of integrity. The specific gravity of the burden of change has moved from genetic to cultural evolution. Genetic evolution itself has taken millions of years; cultural evolution is a child of no more than twenty or thirty thousand years. What will be the future of our evolutionary course? Will it be cyclical in the classical sense? Will it be linear in the modern sense? Certainly, life is more than mere endless repetition. We must restore the importance of each moment, each deed. This is impossible if the future is nothing but a mechanical extrapolation of the past. Dignity becomes possible only with choice. The choice is ours.

In this light, evolution shows man arisen by a creative power inherent in the universe. The immense ancestral effort that has borne man invests him with a cosmic responsibility. Michelangelo's image of Adam created at God's command becomes a more intelligent symbol of man's position in the world than does a description of man as a chance aggregate of atoms or cells. Each successive stage of emergence is more comprehensive, more meaningful, more fulfilling, and more converging, than the last. Yet a higher faculty must always operate through the levels that are below it. The higher faculty must enlist the laws controlling the lower levels in the service of higher principles, and the lower level which enables the higher one to operate through it will always limit the scope of these operations, even menacing them with possible failure. All our higher endeavors must work through our lower forms and are necessarily exposed thereby to corruption. We may thus recognize the cosmic roots of tragedy and our fallible human condition. And language itself as the power of universals, is the basic expression of man's ability to transcend his environment and to transmute tragedy into a moral and spiritual triumph.

This relation of the higher to the lower applies again when an upper level, such as consciousness or freedom, endeavors to reach beyond itself. If no higher level can be accounted for by the operation of a lower level, then no effort of ours can be truly creative in the sense of establishing a higher principle not intrinsic to our initial condition. And establishing such a principle is what all great art, great thought, and great action must aim at. This is indeed how these efforts have built up the heritage in which our lives continue to grow.

Has man's intelligence broken through the limits of his own powers? Yes and no. Inventive efforts can never fully account for their success, but the story of man's evolution testifies to a creative power that goes beyond that which we can account for in ourselves. This power can make us surpass ourselves. We exercise some of it in the simple act of acquiring knowledge and holding it to be true. For, in doing so, we strive for intellectual control over things outside ourselves, in spite of our manifest incapacity to justify this hope. The greatest efforts of the human mind amount to no more than this. All such acts impose an obligation to strive for the ostensibly impossible, representing man's search for the fulfillment of those ideals which, for the moment, seem to be beyond his reach.

The origins of one person can be envisaged by tracing that person's family tree all the way back to the primeval specks of protoplasm in which his first origins lie. The history of the family tree converges with everything that has contributed to the making of a human being. This segment of evolution is on a par with the history of a fertilized egg developing into a mature person, or the history of a plant growing from a seed; it includes everything that caused that person, or that plant, or that animal, or even that star in a galaxy, to come into existence. Natural selection plays no part in the evolution of a single human being. We do not include in the mechanism of growth the possible adversities which did not befall it and hence did not prevent it. The same principle of development holds for the evolution of a single human being; nothing is gained in understanding this evolution by considering the adverse chances which might have prevented it.

In our search for a reasonable cosmic view, we turn in the first place to common understanding. Science largely relies for its subject matter on a common knowledge of things. Concepts of life and death, plant and animal, health and sickness, youth and age, mind and body, machine and technical processes, and other innumerable and equally important things are commonly known. All these concepts apply to complex entities, whose reality is called into question by a theory of knowledge which claims that the entire universe should ultimately be represented in all its aspects by the physical laws governing the inanimate substrate of nature.

Our new theory of knowledge, as the authors in this Series try to demonstrate, rejects this claim and restores our respect for the immense range of common knowledge acquired by our experience of convergence. Starting from here, we sketch out our cosmic perspective by exploring the wider implications of the fact that all knowledge is acquired and possessed by relationship, coalescing, merging.

We identify a person's physiognomy by depending on our awareness of features that we are unable to specify, and this amounts to a convergence in the features of a person for the purpose of comprehending their joint meaning. We are also able to read in the features and behavior of a person the presence of moods, the gleam of intelligence, the response to animals or a sunset or a fugue by Bach; the signs of sanity, human responsibility, and experience. At a lower level, we comprehend by a similar mechanism the body of a person and understand the functions of the physiological mechanism. We know that even physical theories constitute in this way the processes of inanimate nature. Such are the various levels of knowledge acquired and possessed by the experience of convergence.

The authors in this Series grasp the truth that these levels form a hierarchy of comprehensive entities. Inorganic matter is comprehended by physical laws; the mechanism of physiology is built on these laws and enlists them in its service. Then, the intelligent behavior of a person relies on the healthy functions of the body and, finally, moral responsibility relies on the faculties of intelligence directing moral acts.

We realize how the operations of machines, and of mechanisms in general, rely on the laws of physics but cannot be explained, or accounted for, by these laws. In a hierarchic sequence of comprehensive levels, each higher level is related to the levels below it in the same way as the operations of a machine are related to the particulars, obeying the laws of physics. We cannot explain the operations of an upper level in terms of the particulars on which its operations rely. Each higher level of integration represents, in this sense, a higher level of existence, not completely accountable by the levels below it yet including these lower levels implicitly.

In a hierarchic sequence of comprehensive levels each higher level is known to us by relying on our awareness of the particulars on the level below it. We are conscious of each level by internalizing its particulars and mentally performing the integration that constitutes it. This is how all experience, as well as all knowledge, is based on convergence, and this is how the consecutive stages of convergence form a continuous transition from the understanding of the inorganic, the inanimate, to the comprehension of man's moral responsibility and participation in the totality, the organismic whole, of all reality. The sciences of the subject-object relationship thus pass imperceptibly into the metascience of the convergence of the subject and object interrelationship, mutually altering each other. From the minimum of convergence, exercised in a physical observation, we move without a break to the maximum of convergence, which is a total commitment.

"The last of life, for which the first was made, is yet to come." Thus, CONVERGENCE has summoned the world's most concerned thinkers to rediscover the experience of *feeling*, as well as of thought. The convergence of all forms of reality presides over the possible fulfillment of self-awareness—not the isolated, alienated self, but rather the participation in the life process with other lives and other forms of life. Convergence is a cosmic force and may possess liberating powers allowing man to become what he is, capable of freedom, justice, love. Thus man experiences the meaning of grace.

A further aim of this Series is not, nor could it be, to disparage science. The authors themselves are adequate witness to this fact. Actually, in viewing the role of science, one arrives at a much more modest judgment of its function in our whole body of knowledge. Original knowledge was probably not acquired by us in the active sense; most of it must have been given to us in the same mysterious way we received our consciousness. As to content and usefulness, scientific knowledge is an infinitesimal fraction of natural knowledge. Nevertheless, it is knowledge whose structure is endowed with beauty because its abstractions satisfy our urge for specific knowledge much more fully than does natural knowledge, and we are justly proud of scientific knowledge because we can call it our own creation. It teaches us clear thinking, and the extent to which clear thinking helps us to order our sensations is a marvel which fills the mind with ever new and increasing admiration and awe. Science now begins to include the realm of human values, lest even the memory of what it means to be human be forgotten.

No individual destiny can be separated from the destiny of the universe. Alfred North Whitehead has stated that every event, every step or process in the universe, involves both effects from past situations and the anticipation of future potentialities. Basic for this doctrine is the assumption that the course of the universe results from a multiple and never-ending complex of steps developing out of one another. Thus, in spite of all evidence to the contrary, we conclude that there is a continuing and permanent energy of that which is not only man but all of life. For not an atom stirs in matter, organic and inorganic, that does not have its cunning duplicate in mind. And faith in the convergence of life with all its multiple manifestations creates its own verification.

We are concerned in this Series with the unitary structure of all nature. At the beginning, as we see in Hesiod's *Theogony* and in the Book of Genesis, there was a primal unity, a state of fusion in which, later, all elements become separated but then merge again. However, out of this unity there emerge, through separation, parts of opposite elements. These opposites intersect or reunite, in meteoric phenomena or in individual living things. Yet, in spite of the immense diversity of creation, a profound underlying convergence exists in all nature. And the principle of the conservation of energy simply signifies that there is a *something* that remains constant. Whatever fresh notions of the world may be given us by future experiments, we are certain beforehand that something remains unchanged which we may call *energy*. We now do not say that the law of nature springs from the invariability of God, but with that curious mixture of arrogance and humility which scientists have learned to put in place of theological terminology, we say instead that the law of conservation is the physical expression of the elements by which nature makes itself understood by us.

The universe is our home. There is no other universe than the universe of all life including the mind of man, the merging of life with life. Our consciousness is evolving, the primordial principle of the unfolding of that which is implied or contained in all matter and spirit. We ask: Will the central mystery of the cosmos, as well as man's awareness of and participation in it, be unveiled, although forever receding, asymptotically? Shall we perhaps be able to see all things, great and small, glittering with new light and reborn meaning, ancient but now again relevant in an iconic image which is related to our own time and experience?

The cosmic significance of this panorama is revealed when we consider it as the stages of an evolution that has achieved the rise of man and his consciousness. This is the new plateau on which we now stand. It may seem obvious that the succession of changes, sustained through a thousand million years, which have transformed microscopic specks of protoplasm into the human race, has brought forth, in so doing, a higher and altogether novel kind of being, capable of compassion, wonder, beauty and truth, although each form is as precious, as sacred, as the other. The interdependence of everything with everything else in the totality of being includes a participation of nature in history and demands a participation of the universe.

The future brings us nothing, gives us nothing; it is we who in order to build it have to give it everything, our very life. But to be able to give, one has to possess; and we possess no other life, no living sap, than the treasures stored up from the past and digested, assimilated, and created afresh by us. Like all human activities, the law of growth, of evolution, of convergence draws its vigor from a tradition which does not die.

CONVERGENCE is committed to the search for the deeper meanings of science, philosophy, law, morality, history, technology, in fact all the disciplines in a transdisciplinary frame of reference. This Series aims to expose the error in that form of science which creates an unreconcilable dichotomy between the observer and the participant, thereby destroying the uniqueness of each discipline by neutralizing it. For in the end we would know everything but *understand nothing,* not being motivated by concern for any question. This Series further aims to examine relentlessly the ultimate premises on which work in the respective fields of knowledge rest and to break through from these into the universal principles which are the very basis of all specialist information. More concretely, there are issues which wait to be examined in relation to, for example, the philosophical and moral meanings of the models of modern physics, the question of the purely physico-chemical processes versus the postulate of the irreducibility of life in biology. For there is a basic correlation of elements in nature, of which man is a part, which cannot be separated, which compose each other, which converge, and alter each other mutually.

Certain mysteries are now known to us: the mystery, in part, of the universe and the mystery of the mind have been in a sense revealed out of the heart of darkness. Mind and matter, mind and brain, have converged; space, time, and motion are reconciled; man, consciousness, and the universe are reunited since the atom in a star is the same as the atom in man. We are homeward bound because we have accepted our convergence with the Cosmos. We have reconciled observer and participant. For at last we know that time and space are modes by which we think, but not conditions in which we live and have our being. Religion and science meld; reason and feeling merge in mutual respect for each other, nourishing each other, deepening, quickening, and enriching our experiences of the life process. We have heeded the haunting voice in the whirlwind.

The Möbius Strip

The symbol found on jacket and binding of each volume in Convergence is the visual image of *convergence*—the subject of this Series. It is a mathematical mystery deriving its name from Augustus Möbius, a German mathematician who lived from 1790 to 1868. The topological problem still remains unsolved mathematically.

The Möbius Strip has only one continuous surface, in contrast to a cylindrical strip, which has two surfaces—the inside and the outside. An examination will reveal that the Strip, having one continuous edge, produces *one* ring, twice the circumference of the original Strip with one half of a twist in it, which eventually *converges with itself.*

Since the middle of the last century, mathematicians have increasingly refused to accept a "solution" to a mathematical problem as "obviously true," for the "solution" often then becomes the problem. For example, it is certainly obvious that every piece of paper has two sides in the sense that an insect crawling on one side could not reach the other side without passing around an edge or boring a hole through the paper. Obvious—but false!

The Möbius Strip, in fact, presents only one mono-dimensional, continuous ring having no inside, no outside, no beginning, no end. Converging with itself it symbolizes the structural kinship, the intimate relationship between subject and object, matter and energy, demonstrating the error of any attempt to bifurcate the observer and participant, the universe and man, into two or more systems of reality. All, all is unity.

I am indebted to Fay Zetlin, Artist-in-Residence at Old Dominion University in Virginia, who sensed the principle of convergence, of emergent transcendence, in the analogue of the Möbius Strip. This symbol may be said to crystallize my own continuing and expanding explorations into the unitary structure of all reality. Fay Zetlin's drawing of the Möbius Strip constitutes the visual image of this effort to emphasize the experience of coalescence.

R.N.A.

Introduction

In his study of the concept of civilization *Adventures of Ideas,* A. N. Whitehead wrote: "In each age of the world distinguished by high activity there will be found at its culmination, and among the agencies leading to that culmination, some profound cosmological outlook, implicitly accepted, impressing its own type upon the current springs of action. The ultimate cosmology is only partly expressed, and the details of such expression issue into derivative specialised questions of violent controversy."[1] It is doubtful whether any other conceptual development in the history of civilization has crossed and recrossed so many frontiers of human knowledge. Cosmology has always emerged holding a delicate balance between observation of the natural world, philosophical disputation, belief in the validity of theological dogma, and the perennial attitude of man that, however far his vision has been extended, the human being occupies a privileged position in the universe.

The earliest surviving evidence of cosmological thought comes from Egypt and Mesopotamia in the middle of the third millennium B.C. These early cosmologies embodied principles considered necessary to establish the prestige of the God re-

1. Cambridge, England: Cambridge University Press, 1933, pp. 13–14.

sponsible for the Creation, and for the survival of the civilized state. The earliest surviving maps from the Babylonian period exhibit this primitive cosmological outlook. The Earth is shown as a flat disk floating on the ocean with Babylon as its center. Mythology, astrology, and religion remained dominant influences on cosmology for two millennia after these earliest records. Then, with the rise of Greece we see the influence on cosmology of the genius for rational inquiry. As early as the ninth century B.C., the new view of the world is exhibited by Homer. In the *Iliad* we are informed that on the great shield of Achilles was wrought "the earth, therein the heavens, therein the sea, and the unwearied sun, and the moon at full, and therein all the constellations wherewith heaven is crowned." A few centuries later, and particularly with the Milesian philosophers of the fifth and sixth centuries B.C., the observation of the world becomes a major influence on the emergence of cosmology. A most significant indicator of this radical transformation is the famous prediction by Thales of the solar eclipse of 585 B.C. and his statement that the Moon was illuminated by the Sun. From his younger contemporary, Anaximander, we find the first evidence for the belief that the Earth was curved. Although Anaximenes lived after Anaximander and returned to a belief in the flat Earth, he was the first to distinguish between planets and stars. Soon the question whether the Earth was flat, curved like a cylinder, or a sphere was no longer a matter for disputation, because in the third century B.C. Eratosthenes *calculated* the circumference of the Earth from measurements of the shadow cast by a vertical gnomon. Moreover, his calculations were within a few percentage points of the correct figure.

Starting from this epoch, when opinions about the world and the universe were derived from mensuration, I have attempted to trace the critical factors in the emergence of cosmology.[2] The fascinating interplay across the frontiers of

2. For an account of earlier cosmological beliefs dating back to the third millennium B.C., see S. G. F. Brandon, *Creation Legends of the Ancient Near East* (London, 1963). See also *Ancient and Medieval Science from the Beginnings to 1450,* ed. René Taton (Paris, 1957; English translation, London, 1963).

knowledge has been continuous and unending. The stabiliza-
tion of the transcendental and scientific approaches to the cosmological problem has been a vital issue in the emergence of
civilization. At no epoch in the last two millennia has either
approach been dominant, and neither can claim ultimate authority today. For two millennia the advance of cosmology has
illuminated the injunction of Francis Bacon: "And generally
let this be a rule, that all partitions of knowledges be accepted
rather for lines and veins, than for sections and separations;
and that the continuance and entireness of knowledge be preserved. For the contrary hereof hath made particular sciences
to become barren, shallow and erroneous; while they have not
been nourished and maintained from the common fountain."[3]

3. Francis Bacon, *The Advancement of Learning*, book 2: *The Proficience and Advancement of Learning Divine and Human* (London, 1605).

1

The Geocentric Cosmology

From my home in the Cheshire countryside I have two dramatic views. When I awake in the morning I see the Sun rising in the east over Mow Cop, the hill on which, two hundred years ago, a follower of John Wesley preached in the open air to a congregation of thousands. Although Wesley lived two hundred years after the publication of the Copernican hypothesis that the Earth was in motion around the Sun, he was reluctant to accept the proposition, declaring that the new ideas "tend toward infidelity." He, and the many others who resisted Copernican ideas, had both the Scriptures and common sense on their side. It is the Sun that I see climbing the heavens above Mow Cop; I neither see nor feel the rotation of the Earth. On the other hand, I live and have been educated in the twentieth century. My teachers never questioned that the Earth was in motion around the Sun, because the diurnal, seasonal, and annual changes we witness in the heavens can be explained more simply and accurately by this supposition than by the complexities of models based on the idea that the Earth is fixed and motionless at the center of the stellar sphere.

Twenty years ago the Soviets launched the first Sputnik, and since that time many hundreds of satellites and space probes have been launched from Earth. These events have provided a clearer demonstration than ever before of the diurnal rotation of the Earth. Our television programs from other parts of the world come to us via satellites placed in geostationary orbits, that is, satellites in orbits 22,300 miles above the surface of the Earth so that their orbital period is the same as the period of rotation of the Earth. Unlike satellites in orbits at lower heights, which can be seen moving across the sky, communication satellites appear to be stationary above a fixed point on the Earth. No one seriously contends today that the Earth is fixed in space, but, indeed, that has been the belief for most of the history of the civilized world. The transition of the concept from a fixed, motionless Earth at the center of the universe to the idea of an Earth in motion around the Sun marks one of the most critical and tortuous evolutionary changes in cosmology.

If I turn away from the sunrise over Mow Cop, I see from my home an altogether different view—the white circular bowl of the great radio telescope at Jodrell Bank. It is a mark and a symbol of the twentieth-century view of cosmology, for it is receiving signals from objects so far away in time and space that they originated many billions of years ago, when the universe was very young. At least, we believe today that this is the case. Our belief arises from the second of the major evolutionary changes in cosmology occurring in the first three decades of this century.

We learned that the Sun and the planets could no longer be regarded as occupying a privileged, central position in the scheme of the cosmos. On the contrary, the Earth was in motion around the Sun; an average star situated in space far from the central regions of the Milky Way system of a hundred billion stars. We learned also that this system of stars was localized in space; it was one galaxy among millions apparently uniformly distributed throughout time and space. The evolutionary change during those years was marked also by a radical change in our view of time and space. The absolute time and space of Newton and the ancients, which existed independently of matter, had to be abandoned in favor of the concepts of general relativity—an evolutionary change in cosmology of such magnitude that the universe could no longer be viewed as static and unchanging. Cosmology evolved at this stage into concepts of the evolutionary universe, a universe that apparently had its beginning many billions of years ago, and over this vast time span had been expanding from a condition of extreme density and temperature. The discovery in 1965 of the relic radiation from the early universe seems to leave little doubt as to the correctness of this view.

My alternating view of Mow Cop and the radio telescope inspires caution. John Wesley's tenacious belief in the Scriptures led to his caution about the Copernican doctrine, but he embraced the contemporary view of comets as natural astronomical phenomena and calmed the hysteria about the supposed evil terrestrial occurrences associated with their appearance in the heavens. If he lived today he would surely similarly embrace the modern astronomical view in a great unitary synthesis of theological and scientific knowledge, as Thomas Aquinas had done five hundred years before him. Aquinas's struggle was with the ancient physical system of Aristotle. Wesley's struggle was with the universe revealed by Copernicus and Newton. Today we stand in awe at the vastness of the universe and of a cosmology that has evolved beyond the theological conflicts of past millennia to question the very concepts of human comprehension of the natural world.

The Universe of Aristotle

The writings of Aristotle in the decade from about 335 B.C. until his death exerted a dominant influence on physics and astronomy for nearly two thousand years. Compared with the rapid change that has occurred from the sixteenth and seventeenth centuries to the present, this influence on cosmology may well turn out to be the most lasting in the history of the world. In one important respect Aristotle's universe was similar to that of the twentieth century: it was self-contained and self-sufficient with no place, nothing, outside itself, and his logic denied the possibility of the existence of a void. In Aristotle's universe matter and space were entwined; they were aspects of the same feature of the natural world as they are in the Einstein universe of general relativity. This aspect of Aristotle's teaching is purely fortuitous in relation to modern cosmology, for in nearly all respects his cosmology was disastrous. He wrecked the promising developments of the Ionian school and stifled the emergence of heliocentric theory for over a thousand years.

In the sixth century B.C. the Pythagoreans maintained that the planets were eternal and divine. From this they concluded that the planets' motions must be uniform and circular. In the next century Plato enunciated the contrary argument that the motions of the planets were uniform and circular and hence were governed by a divine soul. Aristotle absorbed this Platonic doctrine that the stars were divine beings, changeless and eternal. For Aristotle the heavens were a sphere because the sphere is a perfect figure; only circles have no beginnings and no ends, and hence, to be eternal, the heavenly bodies must move in perfect circles. The logical argument then proceeds to establish that since the center of a rotating body is at rest, therefore the Earth must be at rest in the center of the universe. On the question of the elements, Aristotle argued that Earth was cold by nature and therefore moved downward; this had to be balanced by fire, which was hot by nature and moved upward. He accounted for water and air by the argument, derived from Plato, that the three dimensions of solid bodies correspond to cube numbers, which require two means to unite them (e.g., 1:2:4:8). Aristotle still needed to account for circular motion: fire and air moved upward, earth and water downward. He therefore introduced a fifth element, the ether, which was itself eternal, not subject to change and moving with the perfection of the circular motion. The heavenly bodies, being perfect, were made of this ether. This was a grave retrogression from the standpoint of the Ionians who, correctly, believed that the heavens were of the same stuff as the Earth and, like the Earth, were subject to change and decay.

It is ironic that the mathematical achievements of the Academy enabled Aristotle to construct a cosmological model on this basis. Although the stars moved in an orderly circular manner, it was difficult to account for the apparent irregular motion of the planets across the sky. Eudoxus had found an approximate solution by analyzing the planetary paths into more than thirty circular motions. His scheme was further improved by Callippus (a contemporary of Aristotle), who introduced even more spheres of revolution. Aristotle simply transformed this mathematical scheme into a mechanical model of the universe by assuming that outside the terrestrial sphere (of earth, water, fire, and air) there were fifty-five homocentric celestial spheres carrying with them the heavenly bodies as they revolved around the Earth. The outermost sphere was that of the fixed stars, and the outer surface of this sphere defined the end of the universe. The spheres and the heavenly bodies were composed of the changeless and indestructible ether.

Aristotle's cosmological scheme was based on irrefutable logical deductions from false premises. He had maintained a belief in the perfection of the motion and constitution of divine heavenly bodies; only the material of the universe inside the sublunary sphere was subject to change and decay. Furthermore, he derived a working model for the universe in accordance with the most advanced mathematical deductions of the Academy concerning the apparently irregular motion of the planets across the sky. All the spheres were assumed to be in contact and were driven by the motion of the outer sphere of the fixed stars until this motion, transmitted sphere by sphere, drove the planets and eventually the lowest sphere, which carried the Moon. God was the Prime Mover of the outer sphere, but the means by which He thus animated a corporeal sphere proved a difficulty in the completion of the

Aristotelian logical scheme for the universe. Aristotle simply explained that it was God's desire that set the outermost sphere in motion.

The Rejection of the Heliocentric Hypothesis

Soon after the death of Aristotle, the preeminence of Athens as a center of research passed to Alexandria. But before this occurred, Heraclides of Pontus, who survived Aristotle by only a few years, suggested that the motion of the planets Venus and Mercury across the heavens could be accurately described on the assumption that they revolved around the Sun and not the Earth. He also decided that the motion of the stars could be explained as a consequence of the rotation of the Earth rather than by the rotation of the stellar sphere around a fixed Earth. This concept of Heraclides, namely, that the Earth rotated on its axis once in twenty-four hours, was a major advance in cosmological thought that no one before him seems to have suggested as the reason for the apparent motion of the stars.

Aristarchus of Samos was born within a few years of the death of Heraclides. He was a contemporary of Euclid, and both added renown to the new Alexandrian school. It is from the writings of Archimedes, the younger contemporary of Aristarchus, that we are made aware of his hypothesis extending the theories of Heraclides to embrace the movement of all the planets, including the Earth, around the Sun. The work containing the arguments and hypotheses of Aristarchus is lost, but evidence for his views is contained in a letter written by Archimedes to the king of Syracuse in which, referring to Aristarchus, he writes: "His hypothesis is that the fixed stars and the Sun remain unmoved, and that the Earth revolves about the Sun in the circumference of a circle, the Sun lying in the middle of its orbit."

Thus, in the third century B.C., and in defiance of the Aristotelian doctrine, the heliocentric idea had been correctly formulated. The theory was restated by Seleucus in the next century. Seleucus, a Babylonian astronomer who adopted a Macedonian name and published in Greek, also formulated correct theories about tides. After Aristarchus and Seleucus, this correct heliocentric idea was abandoned for nearly two thousand years.

There were several reasons for this rejection. The idea was an affront to the religious prejudices of the age. According to Plutarch, Cleanthes, a contemporary of Aristarchus, thought it the duty of the Greeks to indict Aristarchus on a charge of impiety for "putting in motion the Hearth of the Universe."[1] Probably neither these prejudices nor the strength of Aristotle's teaching would have destroyed the heliocentric theory had it not been for the influence of Hipparchus, who lived from 161 to 126 B.C. Certainly Hipparchus was one of the great astronomers of antiquity, and it is ironic that his influence was predominant in the rejection of heliocentric motion in favor of the idea of a fixed Earth. The eminence of Hipparchus is beyond question; he wrote the first systematic treatise on trigonometry, he discovered the precession of the equinoxes, he obtained a value for the length of the lunar month that was correct to one second, and he catalogued 850 stars. The skill of Hipparchus may be assessed from the remarkable nature of his discovery of the precession of the equinoxes. He had at his disposal the ancient records of the Babylonian astronomers and the results of 150 years of observations at the Observatory of Alexandria, and from these he concluded that the longitudes of the stars increased by 50 seconds of arc per year. The same records that led Hipparchus to the discovery of precession also contained far more information about the apparent motion of the planets across the sky than had been available to any previous astronomer. He decided that the sys-

1. *De facie in orbe lunae*, chapter 6, in vol. 12 of the Loeb translation of Plutarch's *Moralia* (1957). The passage is also quoted in T. L. Heath, *Aristarchus of Samos* (Oxford, 1913). I am indebted to George B. Kerferd, professor of Greek at the University of Manchester, for informing me that Cleanthes wrote a treatise *Against Aristarchus* that is no longer extant.

tem of Aristarchus, which proposed that the Earth and the planets moved in circles around the Sun, failed to account for the apparent irregularities observed in planetary positions. It must be remembered that in the final victory of the heliocentric hypothesis, seventeen hundred years later, the same difficulties arose concerning the prediction of planetary positions based on simple circular motions—difficulties not surmounted until the realization by Kepler that the motions were not circular but elliptical. In any case, the effect on Hipparchus of this failure of the heliocentric concept was to turn his attention to the alternative geocentric scheme involving the fixed Earth.

The Problem of Planetary Motions

Observers of antiquity were able to record the apparent motion of the Sun against the sphere of fixed stars simply by recording the changing appearance of constellations near dawn and dusk. Changes in the Sun's position in the heavens with respect to the observer on Earth could readily be measured by the gnomon's shadow. The earliest quantitative cosmological schemes accounted for these somewhat complex changes by introducing the idea that two simple circular motions were involved. With the Earth fixed at the center of the universe, the changing position of the Sun with respect to the observer, and its annual motion against the stars, could be satisfactorily explained as follows:

1. A great sphere carrying the stars rotated westward about a fixed axis once every 23 hours 56 minutes.
2. The Sun moved eastward, completing a circular motion once every 365.25 days. The circle the Sun's motion defined on the stellar sphere was tilted 23.5 degrees to the equator of the stellar sphere.

This cosmology of the two-sphere universe might well have survived without contention until the invention of the telescope had it not been for problems in accounting for the apparent motion of planets. Seven planets were recognized in antiquity. With the Sun and the Moon in this category, the other five were Mercury, Venus, Mars, Jupiter, and Saturn. The planets exhibit a diurnal westward motion with the stars and move gradually eastward among the stars. They are always seen in a strip of sky extending from about 8 degrees on either side of the ecliptic (i.e., the great circle on the stellar sphere projected by the motion of the Sun). The eastward motion, which eventually gives the planets a complete circuit of the ecliptic, was found to be about a year for Mercury and Venus, 687 days for Mars, 12 years for Jupiter, and 29 years for Saturn. Whereas the motion of the Sun was uniform around the ecliptic, however, this was observed not to be true of planets. Their eastward progression was interrupted by intervals of westward, or retrograde, motion. These brief intervals of retrograde motion occurred approximately every 116 days for Mercury, 584 days for Venus, 780 days for Mars, 399 days for Jupiter, and 378 days for Saturn. There was further complication: although the five planets showed this regular retrograde motion, another characteristic distinguished Mercury and Venus from the remainder. The direct and retrograde motions of Mercury and Venus had the effect of maintaining their positions fairly close to the Sun. Mercury is always found within 28 degrees of the Sun and Venus within 45 degrees (this angle is known as the elongation of the planet). This shuttle effect of the apparent motion of Mercury and Venus across the Sun means that their appearances in the sky are either as "morning stars" or "evening stars" depending on whether they are to the west or east of the Sun. Only when the ancient observers began to analyze the motions of these planets did they realize that the same planets appeared sometimes in the morning sky and sometimes in the evening sky.

The behavior of the other three planets—Mars, Saturn, and Jupiter—was observed to be different from that of Mercury and Venus. Relative to the Sun they could be observed in any part of the ecliptic: sometimes close to or in conjunction with the Sun, 180 degrees along the ecliptic from the Sun (i.e., in opposition), or in intermediate positions. The retrograde motion was observed to occur only when they were in opposition, and at such times these planets were at their brightest."

The problem facing the astronomers was not merely to provide a qualitative account of these apparently complex motions of planets across the sky but also a quantitative one of predicting precisely where in the sky planets would appear at any time. If irregularities in motion are neglected, then the homocentric sphere concept embodied in Aristotle's cosmology can be made to account for the various periods of the planets. The problem first seems to have been posed by Plato, early in the fourth century B.C. Plato asked for a solution involving uniform and ordered movements that could account for the apparent movement of planets. His pupil Eudoxus devised an ingenious geometrical scheme for the homocentric sphere model, which could account reasonably well for the motion of the Sun and Moon and the regular motion of planets, but failed to account for retrogressions in their motion.

The Theory of Epicycles

It was precisely this difficulty that led to the emergence of the theory of epicycles and deferents as the dominant cosmology until the sixteenth century. When Hipparchus found that the simple heliocentric theory of Aristarchus could not account for the peculiarities of planetary motion, he turned to the theory of epicycles developed by Apollonius in the previous century (about 220 B.C.).[2] In this theory the Earth is fix-

2. There is no complete agreement that Apollonius was the originator of the theory of epicycles. B. L. Van der Waerden, for example, argues that in the sixth century B.C. the Pythagoreans had a geocentric theory of planetary motion based on the assumption of epicycles and eccentric circles that was good for the Sun, Mercury, and Venus but that failed for the other planets. He argues that Plato was well aware of this theory and that when he asked his famous question in the *Dialogues* "By what assumptions of uniform, ordered, circular motion can one save the appearances?" he was posing the problem of *all* planets. *Jr. Hist. Astr.* 5(1974): 175. The homocentric solution proposed by Eudoxus in answer to this question was not inferior to that of the Pythagorean epicyclic theory when all the known planets were taken into consideration. I am indebted to George B. Kerferd for the view that the balance of opinion

ed at the center of a rotating circle (the *deferent*). A point on this rotating circle is the center of another rotating small circle, the *epicycle,* and the planet is located on this epicycle. With the sphere of the stars rotating once per day with respect to the fixed Earth, and with suitable periods and directions of motion for the deferent and epicycle, an apparently looped motion for the planet when projected onto the plane of the ecliptic is obtained. This brilliant device, involving only three perfectly regular circular motions, thus easily explains the occasional apparent retrograde motion of the planet. Each planet requires a separate epicycle-deferent system, although with the Sun and Moon no epicycles are needed because they do not exhibit retrograde motion. One deferent for the Sun turning once a year and one deferent for the Moon turning once in 27 ⅓ days are sufficient. For the other five planets, each with its epicycle-deferent system, appropriate variations in sizes and speeds of the epicycle and deferent can be made to give a qualitative fit to the immense variety of observed planetary motions.

Although the simple epicycle-deferent system could be used to explain major observed irregularities in planetary motion, a number of minor difficulties remained. A simple example concerns the apparent motion of the Sun. A straightforward circular deferent with the Earth fixed at the center predicts equal times between the equinoxes. Yet it was observed that the Sun takes six days longer to move along the ecliptic from the vernal to the autumnal equinox than it does to move during the winter from the autumnal to the vernal equinox. Hipparchus devised additions to the epicycle-deferent model that enabled the first reasonably quantitative account to be given of the observed motions and positions of the Sun and the Moon. He introduced two devices into the epicyclic scheme. If the Sun moves in a small circle (a minor epicycle) whose center moves on the circular deferent, then, by adjusting the rates of motion of the minor epicycle and the deferent, many vari-

among modern scholars is that the theory of epicycles was unknown to Plato, Eudoxus, and Aristotle and that although the concept of epicycles may not have originated with Apollonius, it may have had an anonymous origin soon after Eudoxus.

ations of apparent motion can be obtained and the regularity of circular motion is also maintained. For example, in the case of the Sun, if the minor epicycle rotates westward twice while the deferent moves once eastward, then the Sun (situated on the minor epicycle) appears to move in a flattened circle and the apparent irregularity of its motion between the equinoxes can be explained. The other feature introduced by Hipparchus was that of the *eccentric*—a simple deferent whose center is displaced from the center of the Earth. If the center of the eccentric is displaced from the Earth by about three hundredths of the radius of the eccentric, then the extra six days that the Sun spends between the vernal and autumnal equinoxes can be explained.

The epicyclic theory developed by Hipparchus, including possible variations of the devices of the minor epicycle and eccentric, accounted for apparent major irregularities of planetary motion and also enabled predictions to be made about the position of the planets in the sky at given times. In the second century A.D., Ptolemy (Claudius Ptolemaeus of Alexandria) invented one further device in an attempt to make a better quantitative fit between the theory and the observed position of the planets. This was the *equant,* a point displaced from the geometrical center of the circle. The planet was assumed to move in a circle about the geometrical center, but its rate of rotation was uniform with respect to the equant, not with respect to the center of the circle. Thus, when observed from the Earth, at the geometrical center, the planet would appear to be moving at an irregular rate.

Ptolemy's construction of the epicycle theory, using all these various devices of minor epicycle, eccentrics, and equants, is the greatest achievement of ancient astronomy. His *Almagest* is based on the assumption that "the astronomer must strive to demonstrate that all the phenomena in the sky are produced by uniform and circular motions."[3] He first de-

---

3. The original (Greek) title of the work was *Mathematical Syntaxis,* subsequently designated by the appellation *Great Syntaxis.* The work is commonly known from the Arab title (*Al Magisti*) as *The Almagest.*

scribed the complex construction of epicycles and deferents with these various additions that explained the apparent irregularities in motion and also made quantitative predictions for the positions of all the planets. Ptolemy's construction survived for thirteen centuries. Throughout that period no basic changes were made, nor were any new ideas introduced. His involved combinations were made even more complex by astronomers in the Moslem world and in Europe, but only by the further addition of epicycles or known devices of eccentrics and equants in attempts to improve the agreement between predictions of the theory and observed positions of the planets in the sky.

2

The Strength
of the Ancient Cosmology

The geocentric theory that the Earth was placed in the center of the universe was the foundation of a great edifice of unified knowledge into which the Scriptures, the physics of Aristotle, and God and the angels were intricately woven. Astronomy tended to be regarded as a futile subject because, on the authority of the New Testament, there was soon to be a new heaven and a new Earth. Toward the end of the fourth century even St. Augustine enquired, "What concern is it to me whether the Heavens as a sphere inclose the Earth in the middle of the world or overhang it on either side?"[1] But the heavens and the Earth continued to exist, and over the centuries an immense structure of sacred and physical knowledge was woven around the idea of the fixed Earth. In the twelfth century Peter Lombard, a famous theologian at the University of Paris, issued his collection of *Sentences,* which formed a manual of theology for several hundred years. About man's relation to the universe he wrote: "Just as man is made for the sake of God—that is, that he may serve Him—so the universe is made for the sake of man—that is, that it may serve *him*; therefore is man placed at the middle point of the universe,

1. *De Genesi contra Manichaeos,* composed about A.D. 390.

that he may both serve and be served."

In the thirteenth century Thomas Aquinas exerted an immense influence on contemporary thought. He created a synthesis of knowledge of fundamental and lasting importance to the Western world. Aristotelian physics and cosmology and the epicyclic cosmology of Ptolemy had been sporadically attacked by the Christian Church for centuries on the grounds that the science was in conflict with the Scriptures. In 1210 the teaching of Aristotle's physics was prohibited by a Provincial Council in Paris; five years later, the Fourth Lateran Council issued an edict against Aristotelian views. With the contemporary spread of the knowledge of this physics and cosmology into the Western world it is possible that the Aristotelian heritage, together with the epicyclic theories, would have fallen into decay had it not been for Aquinas, who devoted his great intellectual powers to the reconciliation of this science with theological dogma.

In the *Summa theologica* and in his commentary on Aristotle's *Heaven and Earth*, Aquinas resolved the intellectual conflicts of the thirteenth century. The whole vast system of the known universe, the spirituality of man, and the problems of the relations of the whole to God were resolved in a masterly manner on the basis of a scientific description of the world that was almost entirely fallacious. For example, Aristotle's argument for the existence of God was in terms of the First Cause, which itself being unmoved, originated all motion. But Aristotle's cosmology required either forty-seven or fifty-five unmoved movers, and his dialectic leaves the relation of these to God unclear. There was also the vital problem that, according to Aristotle, God was not to be thought of as the Creator of all that exists but rather as One who gave form to that which already existed as primitive matter. But in the fifth century St. Augustine[2] had established as the unshakable foundation of the Christian faith that God created the world from nothing, that He did not create it sooner because time was created

2. *Confessions*, bk. XI, and *The City of God*, bk. XI. *The City of God* occupied Augustine from 413 to 426.

when the world was created. Aquinas resolved such difficulties by arguments which, in that age, seemed irrefutable. For example, regarding the problem of the unmoved Mover and Creator, his argument was that whatever is moved must be moved by something, that an endless regression of this type is impossible, and this leads directly to the proof of one God as the unmoved Mover and Creator. This logical sequence embodies the principle of Aristotle's unmoved Mover yet, unlike Aristotle's dialectic, leads to the Augustinian conclusion. Aquinas fearlessly combined theological reasoning with the scientific beliefs of Aristotle's physics and Ptolemaic cosmology so that the Scriptures appeared as a straightforward account of man, the universe, and the relation of the two to God.

The fearlessness of Aquinas in casting doubt on the literal interpretation of the Scriptures by a sequential logical statement involving Aristotle's cosmology is well illustrated in his treatment of the problem of the Ascension. In *Summa theologica* he proceeds along these lines. In Ephesians 4:10 we read that Christ "ascended up far beyond all heavens, that he might fill all things."[3] About this, Aquinas says that, according to Aristotle, things that are in a state of perfection possess their good without movement. Christ was in a state of perfection; therefore He has this good without movement and therefore it was not fitting for Christ to ascend. Further, Aristotle proves that there is no place *above* the heavens, and since every body occupies a place, Christ could not have ascended. Also, two bodies cannot occupy the same place, and since there can be no passing from place to place except through the middle space, it seems that Christ could not have ascended above all the heavens unless the crystal spheres were divided, which is impossible.

When a logical development of the argument failed,

3. In revised versions, "fill all things" has been translated as "fill the universe." Thus Moffatt's revision reads, "He who ascended above all the heavens to fill the universe," and the Weymouth version (1902) reads, "He who ascended again far above all the Heavens in order to fill the universe."

Aquinas used another stratagem, implying that the biblical passage was intentionally erroneous in order to make it comprehensible to ignorant people. Thus, when considering the passage in Genesis 1:6–7[4] that God made a firmament amid the waters and that the waters under the firmament were divided from the waters above the firmament, Aquinas faced the difficulty of explaining why God did not make the air. His explanation was that Moses was speaking to ignorant people and thus he spoke to them only of things they could feel with their senses. Since water is obviously corporeal, whereas air is not, Aquinas implies that Moses intentionally used water in two senses in this passage, one sense being interpreted by intelligent people as meaning "air."

Aquinas died in A.D. 1274 when Dante Alighieri was nine years old. The works of Aquinas were detailed and erudite; he had reconciled the universe of Aristotle and Ptolemy with the Christian faith. But it was Dante who interpreted the sacred universe for the common people. By allegory and symbolism, through the vast structure of *The Divine Comedy* Dante translates the geocentric cosmology in terms of heaven and hell. His seven-day journey begins on Good Friday in the year 1300. By Saturday he has reached the bottom of hell, at the center of the Earth. Twenty-four hours later he emerges from the long tunnel at sunrise on the other side of the Earth, at the foot of Mount Purgatory. After climbing for three days and nights he reaches earthly paradise at the top of Mount Purgatory and rises successively through transparent spheres. These spheres encompassing the Earth are rotated by angels and each carries one or more of the heavenly bodies with it.[5] The last

---

4. "And God said, Let there be a firmament in the midst of the waters, and let it divide the waters from the waters.

"And God made the firmament, and divided the waters which were under the firmament from the waters which were above the firmament: and it was so."

5. "Then, circle after circle, round
Enring'd each other; till the seventh reach'd
Circumference so ample, that its bow,
Within the span of Juno's messenger,

(Continued on next page)

sphere carried the fixed stars, and above all was the tenth
heaven—the Empyrean, the throne of God.

In this epic work Dante enshrined the Aristotelian and
epicyclic universe, the texts of the Bible, and the theological
reasoning of Aquinas into a seemingly impregnable structure
of beliefs, hopes, and fears that dominated the Western world
for three centuries. Indeed the symbolism of the sacred universe pervaded Western thought long after the ideas of
Aristotle's physics and the epicentric universe had been abandoned. My father, who was a deeply religious man, expressed
the hope to me as a boy that I would not be troubled
throughout my life by the thought and fear of hell, as he had
been. Not troubled perhaps, but nevertheless as intuitively
aware of hell as underneath and heaven above as of the
diurnal movement of the Sun across the sky.

(Continued from page 35)
Had scarce been held entire. Beyond the sev'nth,
Follow'd yet other two. And every one,
As more in number distant from the first,
Was tardier in motion; . . ."

canto 28, *Paradiso.* Cary translation (London and New York, 1889).

3

The Emergence of the
Heliocentric Hypothesis

The system of the universe established in its various aspects by
Aristotle, Ptolemy, Aquinas, and Dante seemed impregnable
at the end of the thirteenth century. Physics and cosmology
gave a common-sense description of observed events, and the
nature of man's relation to God and the universe was
enshrined in the works of Aquinas and Dante. To attack the
fabric of this structure, which seemed at the time to be a final
and satisfactory description of existence, would have been regarded as blasphemy. Nevertheless, the underlying assumptions of its physics and cosmology came under sporadic
criticism over the next two centuries, especially by the Scho-
lastics of the fourteenth century.

In his study of the Copernican revolution, Thomas Kuhn
refers particularly to the arguments advanced by Nicole
Oresme (Nicholas of Oresme, who died in 1382), a member of
the Parisian nominalist school.[1] Oresme subjected Aristotle's
treatise *On the Heavens* to a searching commentary. Although
Oresme finally agreed with Aristotle on every point except his
account of the Creation, Kuhn remarks that "Oresme's
brilliant critique has destroyed many of Aristotle's proofs and

1. Thomas S. Kuhn, *The Copernican Revolution* (Cambridge, Mass.: Harvard University Press, 1957).

suggested important alternatives for a number of Aristotelian positions." In fact, Oresme's teacher, Jean Buridan, had already demolished Aristotle's arguments about the motion of projectiles and had stated the contrary view that "the movement of the stone continually becomes slower until the impetus is so diminished or corrupted that the gravity of the stone wins out over it and moves the stone down to its natural place."[2]

Oresme was much concerned about the question of the movement of the Earth, and his method of analysis is well illustrated in his criticism of Aristotle's refutation of the Pythagorean suggestion that the diurnal motion of the stars could be explained by the assumption that the Earth was rotating on its axis. Oresme draws an analogy between the relative motion of men in two boats and the difficulty of deciding which boat is the moving one. He proceeds: "Therefore I say that if the higher (or celestial) of the two parts of the universe . . . were today moved with a diurnal motion, as it is, while the lower (or terrestrial) part remained at rest, and if tomorrow on the contrary the lower part were moved diurnally while the other part, i.e., the heavens, were at rest, we would be unable to see any change, but everything would seem the same today and tomorrow. It would seem to us throughout that our location was at rest while the other part of the universe moved, just as it seems to a man in a moving boat that the trees outside the boat are in motion."[3] Oresme concludes that nothing about this issue can be decided from the apparent motion of the stars. He maintains that no logical, physical, or scriptural argument can disprove the diurnal rotation of the Earth and that a decision on whether or not the Earth moves must be a matter of faith. Nevertheless, he is at pains to make it clear that he does not believe in the motion of the Earth.

In the next century the German Cardinal Nicholas of Cusa swept aside the entire fabric of the geocentric universe—to-

2. Quoted in ibid., p. 119, from Buridan's *Quaestiones super octo libros physicorum* (Paris, 1509).
3. Quoted in ibid., p. 115, from Oresme's *Le livre du ciel et du monde*.

gether with much other material cherished by the Church. He subjected the *Etymologiae* of Archbishop Isidore of Seville to a searching examination. He investigated the decretals bearing Isidore's name, which formed one of the most valued muniments of the Church. The decretals were held to be the various canons, letters of popes, decrees of councils, and similar documents handed down from the Apostles and collected by Isidore in the eighth century.[4] These documents supported the doctrine, discipline, ceremonials, and many other claims of the Church. Nicholas of Cusa showed that these decretals were a tissue of anachronisms, and in spite of attempts by the Church to suppress his criticism, the decretals were soon recognized as devout but cunning forgeries. With similar disregard for sacred authority Nicholas of Cusa propounded a cosmology in which the Earth was a moving star like the Sun. He maintained that the universe must be an infinite sphere, because nothing smaller would be appropriate to God's omnipotence. Cusa's idea of the universe was a mystical one, making no scientific sense, for he held that each body in the universe was at the same time at the center, on the surface, and in the interior. He declared that the center of the universal sphere coincided everywhere with its periphery.

Thus, a hundred years before Copernicus, the idea of the motion of the Earth had been promulgated anew by an eminent cardinal of the Church. But Cusa's views were based on visions of an infinite deity for whom only an infinite universe would be appropriate, and one who had complete power to resolve the paradoxes thereby created. Only insofar as Cusa's works were widely read and served to accustom man's mind to the apparently absurd idea that the Earth was in motion does his cosmology have significance in the evolution of ideas about the universe.

The real break with the epicyclic Earth-centered universe began in a radically different manner. Today it is often inferred that the break came suddenly and directly with the

---

4. They were probably ninth century French forgeries.

publication in 1543 of the famous *De revolutionibus orbium coelestium* of Copernicus. The belief that Copernicus revived ideas that had arisen sporadically throughout history that the Earth was in motion around the Sun, and thereby severed cosmological thought from the Ptolemaic geometrical concept of the universe, is a loose interpretation of that critical phase in the evolution of cosmology. Copernicus did not introduce the hypothesis that the Earth was in motion around the Sun for iconoclastic or mystical reasons. On the contrary, he was primarily concerned to preserve the sanctity of circular motion, and like others before him, was troubled by the artificial device of the equant, introduced by Ptolemy to improve the agreement of predictions of the epicyclic theory with observations. In the Ptolemaic arrangement, the planet was assumed to move in a circular orbit, with a velocity that was not uniform with respect to the center of the orbit, but to a point (the equant) displaced from the center. When viewed from the Earth at the geometrical center of the orbit, the planet could then be predicted to move at an irregular rate across the heavens—as observed.

Copernicus did not like this scheme. In the Preface to his book addressed to Pope Paul III he wrote:

Those again who have devised eccentric systems, though they appear to have well nigh established the seeming motions by calculations agreeable to their assumptions, have yet made many admissions which seem to violate the first principle of uniformity in motion . . . we find that they have either omitted some indispensable detail or introduced something foreign and wholly irrelevant. This would of a surety not have been so had they followed fixed principles; for if their hypotheses were not misleading, all inferences based thereon might be surely verified. Though my present assertions are obscure, they will be made clear in due course.[5]

5. Quotations in this book from *De revolutionibus* are from the translation by J. F. Dobson and S. Brodetsky in *Occasional Notes of the Royal Astronomical Society* 2, no. 10 (1947).

De revolutionibus was an obscure work, like the *Almagest* of Ptolemy on which it was based. Today we immediately think of all the revolutionary cosmological ideas that were to flow from the Copernican hypothesis, but the primary concern of Copernicus was to explain relatively minor difficulties in the observed motion of the planets. Since the publication of the *Almagest* there had been thirteen centuries of observation of the planets, and the cumulative errors in predicted planetary positions had revealed many discrepancies in the Ptolemaic approach. Referring to earlier ideas about the motion of the Earth, Copernicus wrote:

> I too began to think of the mobility of the Earth; and though the opinion seemed absurd, yet knowing now that others before me had been granted freedom to imagine such circles as they chose to explain the phenomena of the stars. I considered that I also might easily be allowed to try whether, by assuming some motion of the Earth, sounder explanations than theirs for the revolution of the celestial sphere might be so discovered. Thus assuming motions, which in my work I ascribe to the Earth, by long and frequent observations I have at last discovered that, if the motions of the rest of the planets be brought into relation with the circulation of the Earth and be reckoned in proportion to the orbit of each planet, not only do their phenomena presently ensue, but the orders and magnitudes of all stars and spheres, nay the heavens themselves, become so bound together that nothing in any part thereof could be moved from its place without producing confusion of all the other parts and of the universe as a whole.[6]

The motion of the Earth forming the central feature of the hypothesis was not a new idea, as Copernicus was at pains to emphasize in the introduction to *De revolutionibus:*

> I therefore took pains to read again the works of all the

6. Ibid., pp. 4–5.

philosophers on whom I could lay hand to seek out whether any of them had ever supposed that the motions of the spheres were other than those demanded by the mathematical schools. I found first in Cicero that Hicetas [of Syracuse, fifth century B.C.] had realised that the Earth moved. Afterwards I found in Plutarch that certain others had held the like opinion. I think fit here to add Plutarch's own words, to make them accessible to all; "The rest hold the Earth to be stationary, but Philolaus the Pythagorean [fifth century B.C.] says that she moves around the [central] fire on an oblique circle like the Sun and Moon. Heraclides of Pontus and Ecphantus the Pythagorean [fourth century B.C.] also make the Earth to move, not indeed through space but by rotating round her own centre as a wheel on an axle from West to East."[7]

Unable to find any convincing physical justification for the motion of the Earth, Copernicus attempted to justify his revolutionary scheme by emphasizing that not only mathematicians like Aristarchus had envisaged the idea but also that the concept was embedded in the thought of ancient philosophers. But the real significance of the Copernican hypothesis in the evolution of cosmological thought lies not merely in this reintroduction of the idea of the moving Earth but in the complex mathematical scheme he evolved to explain observed planetary motions. By assuming that the Earth and the planets moved in circular orbits around the Sun, Copernicus was able to account for two major anomalies of planetary motion: the retrograde motion and the variation in time required to circle the ecliptic. At least the qualitative agreement of predictions and observations emerged from the theory without the need for the major epicycles and equants of the Ptolemaic theory. With seven circular planetary orbits about the Sun as center, Copernicus was able to produce a mathematically qualitative account of the observed motions. Difficulties re-

7. Ibid.

mained when the theory was applied quantitatively to predict planetary positions. Copernicus solved these problems only by the introduction of eccentrics and minor epicycles. For example, to account for the fact that the Sun apparently moved through the signs of the Zodiac faster during winter than summer he was forced to displace the center of the Earth's orbit from the position of the Sun—and to account for long-term variations, this displaced center had to be in motion. Similar complexities had to be introduced to account for the Moon's motion and that of the planets. For the Moon there were three circular motions involved, and for Mars and the other planets minor epicycles were added. It is a matter of irony that in his aim to simplify the Ptolemaic system by the device of placing the Earth in orbit around the Sun, Copernicus was forced eventually to construct a system of circles, eccentric and minor epicycles, which was neither simpler nor more accurate in its predictions than the Ptolemaic system.

The sequence of events that led Copernicus to this heliocentric hypothesis, and the fact of its survival as a major revolution in human thought, are remarkable features of history. Copernicus was born in 1473 in Torun into a wealthy family of a rich community. The merchants of Torun made great profits by trading mid-European produce with England and Flanders. It is estimated that in the war of 1454–66 against the Teutonic knights, Torun spent more than six times the total annual revenue of Cracow, the capital of Poland. At least some of the wealth of this community was directed to the creation of excellent schools and educational institutions, and although it is uncertain where Copernicus received his early education, he was fortunate in his youthful associations. His uncle, Lukasz Watzenrode (later bishop of Warmia) was rector of St. John's Church, which contained a school that had astronomy as part of its curriculum. At the age of eighteen Copernicus entered the University of Cracow, an academy with two chairs of mathematics and astronomy, one of the chairs having been founded by Marcin Król, a friend and collaborator of the German astronomer Johann Müller (Regiomontanus). At this time Cracow enjoyed a high reputation as a center for astronomical studies, and Copernicus came under the influence of Wojciech of Brudzewo (Brudzewski), who had produced new astronomical tables to fit the meridian of Cracow. In 1493 Hartmann Schedelius wrote that the university "boasts many eminent and very learned men . . . but the science of astronomy stands highest there."[8] Understandably in this environment, during the years when Diaz sailed around the Cape of Good Hope on the way to India and Columbus set out from Palos, Copernicus had his interests awakened in astronomical problems. Probably he first learned about the contradictions in astronomical thought from Brudzewski's writings. But Copernicus was destined for the Church, and having failed to obtain a Warmian canonicate in 1495, he was sent to Bologna to study law.

The influence Regiomontanus exerted on Copernicus, through Brudzewski, in Cracow continued in Bologna where one of his teachers, Domenicco Maria Novara-Anovaria, was also a follower both of Cusa and Regiomontanus. Toward the end of his life (he died three years before the birth of Copernicus) Regiomontanus exhibited increasing discontent with traditional astronomical teaching. Indeed there is evidence that he concluded eventually that the Earth could not be fixed at the center of the universe, but must be in motion, as proposed by Aristarchus. The important point is that Copernicus as a student lived in an age when it was not unusual to consider the possibility that the Earth might be in motion, and it seems probable that his first recorded astronomical observation—that of the Moon in March 1497—was the event that led him to realize that the geocentric theory could not be correct.

Copernicus observed that the disk of the Moon did not vary in size whatever its phase. This confirmed the statement he found in the *Epitome in Almagestum* of Regiomontanus. Yet, ac-

---

8. *Chronicles of the World (Liber chronicorum de historis aetatum mundi, cum descriptione urbium)* (Nuremberg, 1493).

cording to the Ptolemaic theory, elaborated by Regiomontanus in that work, the distance of the Moon from Earth, and hence its apparent size, should have varied by a factor of two from the first and third quarters to the new and full phases.

The return of Copernicus to Poland and the ceremonial and religious demands made on him as the constant companion of Bishop Watzenrode did not suppress his astronomical thought. At the age of thirty-four, in 1507, Copernicus circulated among scholars in Cracow a handwritten document of about twenty pages. This document, commonly known as the *Commentariolus,*[9] was a brief exposition of his concept of the heliocentric theory: "the centre of the Earth is not the centre of the world but only the centre of gravity and of the lunar orbit . . . all heavenly bodies revolve about the Sun which is close to the centre of the world."[10] The *Commentariolus* did not contain the necessary geometrical arguments to support this hypothesis; these were reserved for the larger work *De revolutionibus,* which was to occupy Copernicus and be fitted in among his pastoral duties for the rest of his life.

Copernicus made his own observations of the Sun, the Moon, and the planets to collect evidence for his hypothesis. It is remarkable that he was able to carry out this work with the crude astronomical instruments of the time, as well as to compose *De revolutionibus* among the exhausting professional tasks of his daily life. The translation from the speculations of the *Commentariolus* to the complex geometrical arguments of the *De revolutionibus* occupied Copernicus for twenty-five years after the circulation of the *Commentariolus.* Eleven years later, in March 1543, a printed copy was handed to him as he lay dying. Those twenty-five years in the first half of the sixteenth century when the ideas of the *Commentariolus* were transformed to the heliocentric theory of the *De revolutionibus* mark the epoch when cosmology broke free from the bondage imposed

9. Nicolai Copernici de hypothesibus motuum coelestium a se constitutis commentariolus (*Commentary on the Hypotheses of the Movement of Celestial Orbs,* by Nicolaus Copernicus).

10. Quoted from the extract in J. Szperkowicz, *Nicolaus Copernicus 1473–1973* (Warsaw, 1972).

by the geocentric concept that the Earth must of necessity be fixed at the center of the universe. "So we find underlying this ordination an admirable symmetry in the Universe, and a clear bond of harmony in the motion and magnitude of the orbits such as can be discovered in no other wise."[11]

11. *De revolutionibus*, bk. 1, sect. 10, p. 19.

4

The Survival
of the Copernican Hypothesis

Copernicus restored the idea of the perfection of uniform circular motion at the cost of raising the immense problem of the motion of the Earth. Indeed, the problems and difficulties were so great that it now seems surprising that the heliocentric theory survived in the sixteenth century, whereas all previous hypotheses to this effect had succumbed to the common-sense view that the Earth was stationary. The odds against the survival of the heliocentric theory expounded in *De revolutionibus* were formidable. The arguments in its favor were based on ancient science and concepts; the removal of the need for the artificiality of the equant enabled emphasis to be placed on the aesthetic merits of the theory—the circular symmetries and the "clear bond of harmony in the motion and magnitude of the orbits." In the scales against such support was a mass of astronomical, physical, and theological evidence. Astronomically the theory did not give a greatly superior account of the appearance of the heavens compared with the Ptolemaic theory, and although involving only uniform circular motions, many epicyclic motions were still needed. Physically no tangible evidence could be produced in favor of the motion of the Earth; and in relation to religious dogma the hypothesis was a disaster because the promulgation of the belief in uniform circular motion seemed of little consequence compared with the authority of the Scriptures.

In the absence of the theoretical and observational advances that occurred more than half a century after the death of Copernicus, it seems that the heliocentric idea must inevitably have suffered another long period of eclipse. As it happened, the tortuous history of the manner in which *De revolutionibus* was published and the means by which its revolutionary ideas were disseminated probably secured its survival. In our own age, when the desire to achieve instant priority for ideas often leads to the spread of improperly assimilated results, it is difficult to realize that Copernicus did not seek publication of his life's work on its completion in 1532. We do not know the reason. He was certainly preoccupied with many administrative problems, the warfare with the Teutonic knights continued, and Copernicus was entrusted with far-reaching powers to deal with the reconstruction of Warmia. Years before he completed *De revolutionibus* he had evidence of the opposition to be expected. The news of the hypothesis advanced in the *Commentariolus* had spread throughout Europe, and in a public burlesque near the home of Copernicus in 1525 an immigrant Dutchman had scoffed at the heliocentric concept.

Yet it was an age when progressive thought was encouraged by ecclesiastical patronage. The Vatican's interest in the views of Copernicus is evident from the Preface of *De revolutionibus* where he explained that his

misgivings and actual protests have been overcome by my friends. First among these was Nicolaus Schönberg, Cardinal of Capua, a man renowned in every department of learning. Next was one who loved me well, Tiedemann Giese, Bishop of Kulm, a devoted student of sacred and all other good literature, who often urged and even importuned me to publish this work which I had kept in store not for nine years only, but to a fourth period of nine years. The same request was made to me by many other eminent and learned men. They urged that I should not, on account of my fears, refuse any longer to contribute the fruits of my labours to the common advantage of those interested in mathematics. They insisted that, though my theory of the Earth's movements might at first seem strange, yet it would appear admirable and acceptable when the publication of my elucidatory comments should dispel the mists of paradox.[1]

It seems clear that an explanation of the new theory was given to Pope Clement VII. Of course, this was also the age of the Reformation—Martin Luther's teachings had been officially condemned at the Diet of Worms in 1521. With Luther's insistence that the Bible was the true source of authority, it is hardly surprising that he denounced Copernicus. At about the time when Copernicus was completing *De revolutionibus*, Luther said: "People gave ear to an upstart astrologer who strove to show that the Earth revolves, not the heavens or the firmament, the Sun and the Moon. Whoever wishes to appear clever must devise some new system, which of all systems is of course the very best. This fool wishes to reverse the entire science of astronomy; but sacred Scripture tells us that Joshua[2] commanded the Sun to stand still, and not the earth."[3]

It is the more remarkable that in 1539, from the principal center of the Reformation at Wittenberg, Copernicus received a visitor to whom the world must be forever grateful that *De revolutionibus* was printed. There is irony in the fact that it was on the recommendation of Melanchthon that twenty-two-year-old Georg Joachim von Lauchen (who latinized his name to Rheticus) was appointed professor of mathematics and

---

1. Preface to *De revolutionibus*, pp. 3–4.

2. Joshua 10: 12–13. After Joshua had delivered up the Amorites before the children of Israel, "he said in the sight of Israel, Sun, stand thou still upon Gibeon, and thou Moon, in the valley of Ajalon. And the Sun stood still, and the Moon stayed until the people had avenged themselves upon their enemies."

3. Tischreden in the Walsch ed. of Luther's *Works* (1743). This translation is taken from A. D. White, *A History of the Warfare of Science with Theology in Christendom* (London, 1955) Book I, p. 126. The book was first published in New York in 1896.

astronomy at the University of Wittenberg in 1536. Three
years later he was given leave of absence for the express purpose of visiting Copernicus. He planned to stay with
Copernicus for three weeks and remained for two years. About
this visit of Rheticus to Copernicus, Koestler wrote: "It is a
pity that Rheticus did not report, in his exuberant style, his
first meeting with Canon Copernicus. It was one of the great
encounters of history, and ranks with the meetings of Aristotle
and Alexander, Cortez and Montezuma, Kepler and Tycho
Brahe, Marx and Engels."[4]

Copernicus had finished writing *De revolutionibus,* but no one
knew its contents and he was reluctant to print the work. He
wanted a young disciple who would hand down the essential
contents of the theory to a select few without stirring the emo-
tional response of the masses to the idea of the Earth being in
motion. Rheticus was precisely this disciple, and as he studied
the manuscript his enthusiasm for the work increased. Under
constant pressure from Rheticus and his friend Bishop Giese
of Chelmno, Copernicus first proposed to print the planetary
tables but not the hypothesis of the Earth's motion on which
they were based. Giese objected that such a work would be
"an incomplete gift to the world unless my Teacher
[Copernicus] set forth the reasons for his tables and also included, following the example of Ptolemy, the system or theory and the foundations and proofs upon which he relied."[5]
Eventually a compromise was reached. *De revolutionibus* would
not be printed, but Rheticus would write an account of its
contents, on condition that he nowhere mentioned Copernicus
by name but would refer to the author of the unpublished
manuscript as *domine praeceptor*. In Gdansk in 1540 Rheticus
published this summary under the title *Narratio prima*. The
twenty-six-page manuscript, dedicated to Johannes Schöner,
one of the patrons of Rheticus's visit to Copernicus, gives
evidence of the colossal strain to which Rheticus had been
subjected: "The astronomer who studies the motion of the

4. A. Koestler, *The Sleepwalkers* (London, 1959), pt. 3, p. 155.
5. Quoted in ibid., p. 157, from the *Narratio prima* of Rheticus.

stars is surely like a blind man who, with only a staff
[mathematics] to guide him, must make a great, endless, haz-
ardous journey that winds through innumerable desolate
places. What will be the result? Proceeding anxiously for a
while and groping his way with his staff, he will at some time,
leaning upon it, cry out in despair to Heaven, Earth and all
the Gods to aid him in his misery."[6]

The circulation of the *Narratio prima* among a number of
European scholars increased the pressure for *De revolutionibus*
to be published. At last Copernicus succumbed to the urgings
of Rheticus, who copied and corrected the entire manuscript
from the summer of 1540 to September 1541. In the spring of
1542 Rheticus took the manuscript to Nuremberg to be
printed by Johannes Petreius—an operation that required
constant supervision by Rheticus. Unfortunately he had to
leave Nuremberg in November, before the work was complete,
transferring to a new post—the chair of mathematics in
Leipzig. He left the supervision of the printing to Andreas
Osiander, a leading theologian of the city, who was sympa-
thetic to Copernicus and in whom Rheticus placed complete
trust. This trust was misplaced. Although friendly to
Copernicus, Osiander was a follower of Luther, and having
failed to persuade Copernicus that the work should be presented as a set of hypotheses, not necessarily true, he proceeded to insert an anonymous preface to that effect:

> For it is the duty of an astronomer to compose the history of
> the celestial motions through careful and skillful observa-
> tion. Then turning to the causes of these motions or
> hypotheses about them, he must conceive and devise, since
> he cannot in any way attain to the true causes, such
> hypotheses as, being assumed, enable the motions to be cal-
> culated correctly from the principles of geometry, for the
> future as well as for the past. The present author has per-
> formed both these duties excellently. For these hypotheses

6. Ibid., p. 161.

need not be true nor even probable; if they provide a calculus consistent with the observations, that alone is sufficient.[7]

When *De revolutionibus* appeared, it was doubtful whether this preface was the work of Copernicus. Although Osiander's authorship was discovered and revealed by Kepler in 1609, successive editions of the work continued to print his preface without comment until the edition of 1854 when Osiander's authorship was made clear.

The motives that inspired Osiander to impose himself anonymously in this manner must forever remain a matter of speculation. The most generous opinion must be that as a leader in the Protestant movement he was only too well aware of the antagonisms the heliocentric theory would cause, and that by emphasizing the "hypothetical" nature of *De revolutionibus* he hoped to secure the ultimate acceptance of these revolutionary ideas. At least the anonymous preface justified astronomers of that epoch in using the geometrical concepts of Copernicus without promulgating a belief in the physical motion of the Earth. For a time, at least, those who wished to do so could still believe in the classical doctrine that the science of physics and the geometry of the heavens were separate matters.

If this was the intended strategy of Osiander, his device was not entirely successful. From the Protestants particularly, opposition to the theory of Copernicus was outspoken. The influential academic Philipp Melanchthon, a leader in Luther's reform movement who drafted the *Augsburg Confession* of 1530, attempted to restrain the spread of the heliocentric theory after the publication of *De revolutionibus*. In *Initia doctrinae physicae* (1549) Melanchthon wrote: "The eyes are witnesses that the heavens revolve in the space of twenty-four hours. But certain men, either from the love of novelty, or to make a display of ingenuity, have concluded that the earth moves; and they

7. Quoted from the full text of Osiander's preface, Koestler, *The Sleepwalkers* (London, 1959) p. 565 *n,* as translated by E. Rosen, New York, 1939.

maintain that neither the eighth sphere nor the Sun revolves. . . . Now it is a want of honesty and decency to assert such notions publicly, and the example is pernicious. It is the part of a good mind to accept the truth as revealed by God and to acquiesce in it." Melanchthon cited biblical passages[8] as clearly asserting that the Earth stands still, and added eight other proofs of his proposition that "the earth can be nowhere if not in the centre of the universe."[9]

From Geneva, Calvin published his *Commentary on Genesis* in which he condemned all who departed from the view that the Earth was the center of the universe. For Calvin, as for many others, the Bible was the sole source of authority: "the world also is stablished, that it cannot be moved."[10] On this authority he inquired, "Who will venture to place the authority of Copernicus above that of the Holy Spirit?"[11] Perhaps understandably, scholars were reluctant to own public allegiance to the heliocentric theory. One of the most renowned mathematical and astronomical scholars of the mid-sixteenth century, Peter Apian,[12] adopted an entirely neutral position on the issue. Anxious to maintain the equilibrium of his chair at the University of Ingolstadt, which was under the strictest control of the Church, he neither supposed nor opposed Copernicus and his theory. Apian's influence was tremendous

8. Ecclesiastes 1:4–5. "One generation passeth away, and another generation cometh: but the earth abideth forever. The sun also ariseth, and the sun goeth down, and hasteth to his place where he arose."

9. Translation in White, *History of the Warfare of Science with Theology in Christendom*, p. 127.

10. The first verse of the 93rd Psalm reads: "The Lord reigneth, he is clothed with majesty, the Lord is clothed with strength, wherewith he hath girded himself; the world also is stablished, that it cannot be moved." In the revised versions the last phrase remains identical apart from the replacement of "stablished" by "established."

11. *Commentary on Genesis*, in White, *History of the Warfare of Science with Theology in Christendom*, p. 127.

12. Peter Apianus (Bienewitz) was born in Leissnig in Misnia in 1495. He was professor of mathematics at the University of Ingolstadt and was patronized and ennobled by Charles V. When he died in 1552 he was succeeded by his son, Philip, who was forced to leave his chair when he embraced Protestantism. He was then appointed professor of mathematics and astronomy in Tübingen where in died in 1589.

—Charles V of Germany and Spain[13] was his pupil—and even a plea from him for suspension of judgment on the theory would have calmed the growing antagonism.

Indeed, nowhere is the power of the Protestant movement at that time indicated more forcibly than in the strange case of Rheticus and Reinhold. Rheticus, the devoted pupil of Copernicus to whom we owe the publication of *De revolutionibus,* and his colleague in Wittenberg, Erasmus Reinhold, were both astronomers of the highest order. Reinhold, like Rheticus, was convinced of the correctness of the Copernican theory. Eight years after the publication of the work, Reinhold produced an entirely new set of astronomical tables based on the Copernican theory. The *Prutenic Tables,*[14] the first astronomical predictions to be produced in Europe for three centuries, were greatly superior to any in existence. Any astronomer who used the *Prutenic Tables* implicitly acknowledged the correctness of the Copernican theory, yet Reinhold was not allowed to teach the heliocentric idea. We do not know whether Reinhold could have survived this suppression of his beliefs in Wittenberg (he died in 1553, two years after the issue of the *Prutenic Tables*), or whether, like Rheticus, he would have found freedom to seek and speak the truth elsewhere.[15]

The suppression of the heliocentric concept arose most forcibly from the Protestant movement. An opposition based largely on the presumed contradictions with the Holy Scriptures obscured the more far-reaching influences of the new theory. Seventy years were to elapse before the full implica-

13. Charles V (1500–1558), was king of Spain 1516–56 and emperor 1519–56. At the zenith of his power he ruled over a vast empire of Spain, Germany, The Netherlands, Austria, and Italy. He retired in 1556 to the Hieronymite Monastery of Yuste in Estremadura where he died in 1558.
14. The *Prutenic Tables* were named by Reinhold after his patron, the Duke of Prussia. These tables were used in the reformation of the calendar in 1582 by Gregory XIII.
15. Rheticus moved to his new post in the University of Leipzig during the printing of *De revolutionibus* in 1542. He remained in Leipzig for less than three years, and his subsequent nomadic existence until his death in 1576 is related in Koestler, *The Sleepwalkers,* pt. 3, p. 187.

tions dawned on the Catholic Church, at which time the explosive nature of the antagonism erupted in the suppression of *De revolutionibus* and the trials of Galileo.

5

Tycho Brahe

For the remainder of the sixteenth century, opposition to the heliocentric theory of Copernicus did not proceed entirely from the Church. Tycho Brahe became the preeminent astronomer of the second half of that century. He was a life-long opponent of Copernicanism, and yet, ironically, his astronomical observations led to the final triumph of the heliocentric theory in the first decades of the seventeenth century. Today it is still possible to visit the site where Tycho made his historic series of observations—the island of Hven in the Danish Sound. This small island, once Danish territory, now belongs to Sweden. The road from the small harbor of Baeckviken rises to Uraniebörg where the cupolas of Tycho's observatory have been reconstructed. On this highest point of the island a few houses, the church, and the village school cluster around the historic ground on which Tycho labored four centuries ago. Little remains of his castle except a few foundation stones and a brass plate on which is engraved the ground-floor plan. A statue of Tycho, almost in the school playground, and a small museum complete the scene.

Tycho Brahe was one of the more romantic figures in the history of astronomy. He was born three years after the death

of Copernicus. During his life, Kepler and Galileo rose to the
height of their powers. Tycho was a vital connecting link in
the heroic age of astronomy—the centerpiece between
Leonardo, who was contemporary with Copernicus, and Newton, who was born in 1642, the year of Galileo's death.

Tycho was an extraordinary man—a person of noble birth,
haughty and arrogant, who was at home in the company of
princes. A king gave him an island and unlimited resources to
pursue his astronomical studies. The story of Tycho's up-
bringing reads like a fantasy. His father was the governor of
Helsingborg Castle and his mother was in charge of the
queen's court. But Tycho's illustrious parents were not the
key figures in his early life. He had a distinguished uncle, an
important admiral in the Danish fleet, who had no children
and made his brother promise that, if he had a son, he could
be adopted by his uncle. It is hardly surprising that when
Tycho was born, the father thought better of the bargain and
refused to part with his son. When a second son was born, the
admiral kidnapped the firstborn, Tycho, from the cradle.

The conflict was resolved to the extent that Tycho became
the heir of his uncle, who unfortunately died a few years later
as the result of a naval accident. At age fifteen Tycho was sent
to Leipzig to study law, but before leaving Denmark he had
seen a partial eclipse of the Sun, which inspired him with a
passion for astronomy. At the age of fifty-two, Tycho described this period of his life in his *Astronomiae instauratae mechanica:*

> After I had already in my fatherland Denmark, with the aid
> of a few books, particularly ephemerides, made myself ac-
> quainted with the elements of Astronomy, a subject for
> which I had a natural inclination, now in Leipzig I began to
> study Astronomy more and more. This I did in spite of the
> fact that my governor, who pleading the wishes of my par-
> ents wanted me to study law (which I actually did as far as
> my age allowed it), did not like it and opposed it. I bought
> the astronomical books secretly, and read them in secret in

order that the governor should not become aware of it. . . .[1]

Tycho soon discovered that there were large errors in the existing ephemerides. For example, he found that the predicted time of conjunction of the planets Saturn and Jupiter in 1563 was a month wrong, according to the Alphonsine numbers, and several days wrong, according to the Copernican tables. He made such discoveries with crude equipment, since he was not permitted to buy any instruments.

> I first made use of a rather large pair of compasses as well as I could, placing the vertex close to my eye and directing one of the legs towards the planet to be observed and the other towards some fixed star near it . . . later on, in the year 1564, I secretly had a wooden astronomical radius made. . . . When I had got this radius, I eagerly set about making stellar observations whenever I enjoyed the benefit of a clear sky; and often I stayed awake the whole night through, while my governor slept and knew nothing about it; for I observed the stars through a skylight and entered the observations specially in a small book, which is still in my possession.

Tycho seems to have continued his life in this way at various European universities for a number of years, all the time constructing instruments for the observation of the stars. While studying at Augsburg he became acquainted with the mayor, Paul Hainzel, who was interested in astronomy. The mayor had a considerable property outside the town, and there Tycho built a quadrant of brass and oak eighteen feet in radius and so strongly supported that, although exposed to the wind, it was possible to measure the altitudes of the Sun and the planets to one sixth of a minute of arc, an altogether remarkable astronomical instrument for those days. After

Tycho left Augsburg, the mayor continued to use the quadrant and sent his results to Tycho. Unfortunately the instrument became derelict five years later, and of this event Tycho wrote in later years:

> One could have wished that this excellent instrument had been preserved for a longer period in this place, and had stayed in use, or, else, that another instrument had been constructed in its place. Since however men as a rule are more interested in worldly matters than in things celestial, they usually regard with indifference such happenings which will perhaps be more harmful to them than they themselves realize.

At the age of twenty-six, Tycho considered that his studies in the European universities were complete and returned to Denmark. While in Augsburg he became interested in alchemy, but in that year (1572) an event occurred that led Tycho to abandon everything but astronomy. One of his uncles had built the first paper mill and glassworks in Denmark and encouraged Tycho to continue with his alchemy in these laboratories. On the evening of November 11, 1572, when Tycho was walking home from the laboratories, he noticed a star, as bright as Venus, near the constellation of Cassiopeia, where no bright star had been observed before. Now Tycho had the necessary instruments with which to determine the position of this star, and to the consternation of the leading astronomers of the day, he proved that the new star was motionless among the fixed stars.

Today it is difficult for us to envisage the impact of this discovery on a Europe steeped in the classical Aristotelian doctrine of the fixed, unchanging universe. Tycho continued to observe the star until it faded away eighteen months later. Because his measuring instruments were far superior to any possessed by the professional astronomers, no one was able to question that the event was, indeed, the flare-up of a new star in the firmament. Thus, at twenty-six, Tycho became the

most famous astronomer of his day. The event persuaded him
to abandon his alchemy, which had begun to occupy much of
his time, and thereafter he devoted his life to the study of the
stars.

Tycho had no intention of continuing his work in Denmark.
Indeed, two years later, he had made up his mind to settle in
Basel where he intended to lay the foundations for the revival
of astronomy. His choice fell on Basel because of the healthy
climate and agreeable living, and because it was located at the
point where Italy, France, and Germany met. Thus it would
be easy for him to form friendships with distinguished and
learned men in different places. "I also had the feeling that it
would not be sufficiently easy and convenient for me to pursue
these studies in the fatherland, particularly if I stayed in
Scania and on my property Knudstrup, or in some other
greater province of Denmark where a continuous stream of
noblemen and friends would disturb the scientific work and
impede this kind of study."

But Tycho was not to pursue his destiny in this way; a king
was to intervene. In the *Astronomiae instauratae mechanica* Tycho
describes the sequence of events.

It so happened that while I was inwardly contemplating
these matters and was already making preparations for the
journey, without however revealing my purpose, the noble
and mighty Frederick II, King of Denmark and Norway, of
illustrious memory, sent one of his young noblemen to me at
Knudstrup with a Royal letter bidding me to go to see him
immediately wherever he might be dwelling on Sealand.
When I had presented myself without delay this excellent
King, who cannot be sufficiently praised, of his own accord
and according to his most gracious will offered me that island in the far-famed Danish Sound that our countrymen
call Hven, but which is usually called Venusia in Latin, and
Scarlatina by foreigners. He asked me to erect buildings on
this island, and to construct instruments for astronomical
investigations as well as for chemical studies, and he

graciously promised me that he would abundantly defray the expenses.

After some consideration Tycho accepted this fairy-tale offer, surely the only astronomer ever to have been given dominion over a 2000-acre island, a personal fortune, and unlimited means to construct his buildings and instruments.

Tycho chose the highest site on the island, practically at its center, and in 1576 commenced building the sumptuous castle he called Uraniebörg. The foundation stone was laid by his friend Charles Dançay, the French envoy to Denmark.

When the day approached determined for the laying of the foundation stone, the Excellent Dançay arrived, accompanied by several noblemen besides by some learned men among our common friends to attend this performance, and on the 8th of August in the morning when the rising Sun together with Jupiter was in the heart of Leo, while the Moon was in the western heavens in Aquarius, he laid this stone in the presence of all of us, having first consecrated it with wine of various kinds and praying for good fortune in every respect, in which he was joined by the surrounding friends.

The castle was surrounded by walls 20 feet broad at the base and 22 feet high. These walls delineated a square with sides of 300 feet, but each side of the square was interrupted by a semicircular insertion 90 feet in diameter. At the north and south corners of the wall, Tycho built minor replicas of the main castle, one to house his servants and the other his printing press. The eastern and western corners were formed of great gates built of rough stones, in the Tuscan manner, and on the top of these gates "there were two big English watchdogs in kennels of a suitable size, in order that with their barks they might announce the arrival of people from any direction."

The castle in the center of this enclosure was built in the

form of a square with sides 60 feet long. The building was 75 feet high and included two great round towers, 22 feet in diameter, which contained some of the observing instruments. In addition to this enormous superstructure, a basement and sub-basement contained such things as food stores and alchemical laboratories and furnaces. Tycho said that this part of the house was made up of many rooms and also "required great expenditure for its building, hardly less than the part of the house above ground in the open." It is hardly surprising to read that "after some years it was finished, although in the meantime not a few or insignificant difficulties and delays occurred."

Still Tycho was not content. He continued to manufacture astronomical instruments for which no room could be found on the observatory towers of the castle. About eight years after the construction of the castle had begun, he built another subterranean observatory outside the castle walls. He did this because he wanted to place the most important instruments so that they would not be exposed to the wind. He also wanted to separate his collaborators so that some made observations in the castle and others from this observatory, which he called Stjernebörg, "in order that they should not get in the way of each other or compare their observations before I wanted this."

Tycho worked at Uraniebörg for over twenty years and made a concentrated series of precise observations of the positions and movements of the stars and planets, the like of which had never before been realized. Although Tycho's name is most commonly associated with the supernova observations, without doubt the series of measurements at Uraniebörg represent his great contribution to astronomy. These were the precious treasures from which Kepler evolved his three laws of planetary motion. Tycho Brahe was the great observer; he has no particular claim to greatness as a theorist. On the contrary, his world system was a regression, an uneasy compromise in which he reinstated the Earth as the center of the universe, with the planets circling around the Sun and the Sun around the Earth.

Tycho's mode of life on Hven suited the fabulous nature of his castle. He entertained in a royal manner, with a succession of great banquets to his royal and noble friends, including James VI of Scotland. Alas, Tycho's haughty and arrogant nature eventually led to trouble. He treated his tenants on the island atrociously, and, most fatal, he was rude to anyone he disliked. Unfortunately, this included King Christian IV, who had succeeded Frederick in 1588. Christian IV was unable to turn a blind eye to Tycho's methods of ruling Hven; he wrote letters to Tycho, but they were not answered. After an incident in which Tycho held a family in chains although ordered by the High Court to release them, the great astronomer began to be thoroughly unpopular.

Tycho himself was getting restless after his long stay on Hven, and the restrictions the king managed to place on his huge income confirmed him in his plans for emigration. Tycho's side of the story is restrained. After describing all that he had spent on the island—for example, "more than a tun"[2] of gold on such things as the water works and the paper mill —he says:

From this and several other facts every sensible person will easily conclude that I must have had very weighty reasons, particularly at the age of fifty and with a large family, to leave an island which to me had so great a value, and further my beloved native country and so many relatives and friends I had there. But which and how great reasons have moved me to do so, I prefer not to mention in this place. However, I want to excuse my Serene King, Christian IV, my Most Gracious Lord, who has recently succeeded on the throne of his father, King Frederick, of glorious memory, who laid the foundations of everything there and protected it. For I have no doubt that if he had in time and sufficiently been informed of all the facts of this affair, which cannot but redound to the credit of the realm he would . . . graciously and liberally have preserved these

2. 100,000 rigsdalers. The rigsdaler was a silver coin approximately equivalent in value to an English crown at that time.

studies. . . . For the sake of [astronomy] I have courageous-
ly wanted to bear so many efforts and so much expenditure,
so many disturbances and so much adversity that I have not
even hesitated to leave my native country and everything
that was dearest to me. So great was my desire to investigate the laws of the stars.

The departure from Hven occurred in the spring of 1597.
Tycho took with him his entire suite, library, printing press,
and all his astronomical instruments except some of the larger
ones which followed him later. After various journeyings he
eventually reached Prague in 1599, having secured an appointment with Emperor Rudolf II. Once more Tycho arranged for himself a great castle, outside Prague, and a huge
salary. Although he tackled the conversion of the castle into an
observatory with his usual vigor, his affairs never again ran
smoothly, and he died in 1601 without effectively restarting
his great series of observations. These last restless years were
of enormous significance to astronomy, however, because it
was in the castle near Prague that Tycho Brahe and Kepler
met. It was there that Kepler was at last able to obtain the
precious information that was to be so important to the formulation of the laws of planetary motion.

With the instruments at Uraniebörg Tycho produced a sustained and systematic body of data concerning planetary positions and motions that were without precedent. His records of
planetary positions made at Uraniebörg seem to have been
reliable to about four minutes of arc, and for a fixed star his
positions were accurate to one minute of arc or better. These
accuracies were at least twice as good as any previous measurements. It was the systematic nature of his data, however,
which swept away the ancient records and moreover
eliminated many problems created solely by the inaccuracies
of those data. Neither in the sixteenth century nor today could
any theorist construct a planetary theory capable of reconciling the data available to Copernicus. It is therefore strange
that Tycho failed to grasp the significance of the Copernican

heliocentric concept. He produced many reasons for his rejection of this hypothesis, and developed arguments that sustained many of the classical objections to the idea of an Earth in motion. An extremely powerful argument arose because even with his refined measurements he could not detect any parallax for the planet Saturn. If the Earth moved around the Sun, then the position of the planet against the background of the fixed stars should show a variation throughout the year. Tycho failed to measure any such parallax, and so he argued that if the Earth really was in motion, this negative result could only mean that the distance between the stellar sphere and the planet must be seven hundred times the distance between the planet and the Sun. Such vast scales for the universe were unimaginable (and remained so for another three centuries), and hence Tycho felt forced, for observational reasons, to reject the heliocentric hypothesis. At the same time he did not like the Ptolemaic system and constructed his own Tychonic scheme, which restored the stationary Earth but used another system of epicycles for the planets centered on the Sun. This, Tycho felt, preserved some of the mathematical harmonies of the Copernican system within a geocentric concept.

The Tychonic system was a diversion in the emergence of a cosmology based on the heliocentric concept. His scheme was almost exactly equivalent mathematically to that of Copernicus, and it had the supreme advantage of avoiding the physical and theological conflicts inherent in the heliocentric theory. The system now appears as a peculiar regressive sideline in cosmological development, particularly because, apart from the heliocentric issue, Tycho was involved with two other features of the astronomy of the late sixteenth century that were destructive of Aristotelian cosmology.

To cast doubt on the perfection and immutability of the heavenly sphere was as heretical as suggesting that the Earth was not fixed at the center of the universe. Tycho's observation of the "new star"—which we now recognize as a supernova explosion—on November 11, 1572, was the first unchallengeable evidence that the heavenly sphere was not perfect in its immutability. So far there had been little serious challenge to the Aristotelian edict that change and decay were features only of the sublunary world. Now Tycho, with his superior instruments, could measure the position of this "new star," and he discovered that its position with respect to the fixed stars was unchanging. The new star being clearly visible for eighteen months after its appearance was an apparition recognized by all who studied the heavens. The significance of Tycho's measurements placing it among the fixed stars was immediately recognized as destructive of a belief in the divine perfection of the heavens.

The second feature concerned comets. The observation of a single "supernova" as evidence of the imperfection of the heavens might well have been rejected eventually by those who were determined to prove that Tycho had made an error. Within eighteen months the object of his measurements had become invisible, and no further observations were possible. Unfortunately for those who were determined to preserve the Aristotelian concept of the universe, five years later, on October 27, 1577, Tycho commenced his measurements on a series of comets. During the next twenty years he observed six comets[3] and again failed to measure any parallax that would have confirmed that cometary apparitions belonged to the sublunary sphere.

The discovery that comets had an existence beyond the sublunary sphere touched theological doctrine and popular belief at another sensitive point, apart from the indication that the regions beyond the moon were not perfect.[4] The belief that the

3. November 28, 1580; October 8, 1585; February 8, 1590; July 19, 1593; and July 25, 1596. These and the comet of 1577 are all believed to be long-period comets and have not subsequently been observed. Their periods are probably more than a thousand years.

4. This argument about the nature and origin of comets has a parallel today. Only recently has it been generally accepted by astronomers that all comets are permanently under the gravitational field of the Sun; that is, they are permanent members of the solar system and not occasional visitors from space temporarily moving under the Sun's influence. Some are moving in short-period orbits of a few years; others have long-period orbits and have been seen only on one appearance. There remains uncertainty about the true nature and origin of comets in the solar system.

appearance of a comet was a sign from God, generally a warning to humankind, had its roots deep in history. According to Roman legend, the downfall of Nero was presaged by the appearance of a comet; in the eighth century, Bede declared that "comets portend revolutions of kingdoms, pestilence, war, winds or heat";[5] and Thomas Aquinas accepted and passed on this opinion of Bede. In 1456 a comet appeared at a critical time during the advance of the Turks into Europe.[6] They had captured Constantinople at the time of the apparition. Pope Calixtus III, alarmed at the sight of the comet, decreed several days of prayer "for the averting of the wrath of God, that whatever calamity impended might be turned from the Christians and against the Turks."[7] Papal intercession failed, and the Turks remained in Constantinople. A hundred years later, the powerful Charles V abdicated in the face of fears engendered by the comet of 1556; indeed, there is constant evidence throughout history that people looked on a comet as an object sent by God into the sublunary sphere as a grave warning.

At the time of Tycho's cometary observations the battle between theological and scientific belief in comets reached a peak. In 1577 Jacob Heerbrand of the University of Tübingen was lecturing his students on the moral value of comets. He compared "the Almighty sending a comet, to the judge laying the executioner's sword on the table between himself and the criminal in a court of justice."[8] But perhaps the most remarkable illustration of the conflict occurred in 1577 over the case of Michael Maestlin, a Protestant pastor.[9] He made accurate observations of the 1577 comet, and his results, supporting completely the conclusions of Tycho, led to his ap-

5. See A. D. White, *A History of the Warfare of Science with Theology in Christendom* (London, 1955), p. 175.
6. Subsequently known as Halley's comet. Last seen in 1910, the comet has a period of 75 years and should return to the solar vicinity in 1985. The appearance of this comet for 29 returns has been traced back, using Chinese records, to 239 B.C. by T. Kiang. See "The Past Orbit of Halley's Comet", *Mem. R. Astr. Soc.* 76 (1971): 27.
7. White, *History of the Warfare of Science*, p. 177.
8. Ibid., p. 184.
9. Michael Maestlin was born in Goppingen in 1550. He was appointed professor of mathematics at Heidelberg in 1580 and professor of mathematics at Tübingen in 1584, where he became the teacher of young Kepler. He died in 1631.

pointment as professor of astronomy at Heidelberg. Maestlin
was a Copernican and clearly believed that his observations
had proved the supralunar nature of the comet as an object
obeying natural laws. Yet in his published account of the comet he promulgated the conventional theological view. Gradually, but never completely, the scientific view of comets began
to prevail following Tycho's succession of observations. Nevertheless for another century many professors of astronomy in
Europe, on appointment to their chairs, had to take an oath
that they would not teach that comets were heavenly bodies
obedient to natural law.

Tycho's observations of the supernova of 1572, followed by
his series of observations of comets, cut at the heart of the
long-held Aristotelian belief in the perfection of the heavens.
This was a direct assault on the ancient belief in the divine
nature of the supralunary regions. The second and quite separate feature of his work that was to overthrow the geocentric
view (which he himself supported) was the sustained series of
accurate measurements of the planets, which by the accident
of his self-imposed exile from Hven came into the hands of
Kepler at the turn of the century.

6

The Laws of
Planetary Motion

The years of transition from the sixteenth to the seventeenth century mark a period when, for the first time in history, accurate observation and correct theories were combined in a cosmology in which the idea of the fixed Earth could no longer be sustained. The intellectual synthesis so brilliantly established by Aquinas was destroyed. With no coherent body of physical knowledge to replace Aristotelian physics, and no common-sense view of the world, the paths of science and theology diverged. In the words of John Donne, " 'Tis all in peeces, all cohaerence gone."[1]

The theory and observation were nearly complete in Tycho's work. Oddly, his rejection of Copernicus arose from the precision of his observations. Undoubtedly his intellectual powers were so great that his scientific judgment would have overruled his aesthetic objections to the heliocentric theory had he not placed the wrong interpretation on his failure to determine any measurable parallax for the planet Saturn. Since another two hundred years were to elapse before people had the remotest idea of the relative distances involved between the Sun, planets, and stars, it is hardly surprising that

1. John Donne, *An Anatomie of the World: The First Anniversary* (1611).

Tycho failed to grasp that his failures to measure parallax were not *scientific* arguments against the heliocentric theory. In any case, within a few years of Tycho's death both theory and observation produced unassailable scientific evidence in favor of the Copernican hypothesis. The abruptness of the final dismissal of the geocentric ideas occurred because theory and observation combined at the same time to produce the essential evidence, although in retrospect it seems likely that either alone would have sufficed to lead cosmology away from the bondage of the geocentric concept.

The theoretical problem was resolved by Johannes Kepler. Born in 1571, he seems not to have doubted the views of Copernicus since he first learned about them in his student days at Tübingen. Nevertheless, the fact that he came under the influence of a fervent believer in the heliocentric concept is indicative of the tortuous intellectual paths of that epoch. Kepler's teacher was Michael Maestlin, who was educated at Tübingen by Philip Apian. Maestlin's observations of the 1577 comet, supporting those of Tycho, led to his appointment as professor of astronomy at Heidelberg, which has already been described in chapter 5. Maestlin was intellectually dishonest. He wrapped the ancient theological beliefs concerning comets around his observations, which showed that comets were supralunar phenomena. This humiliating feature of his treatise saved him in the eyes of the Protestants, and when Philip Apian was driven from Tübingen because he refused to sign the Lutheran *Book of Concord,* Maestlin was appointed to his chair at Tübingen.[2]

Notwithstanding his specious Protestant orthodoxy, Maestlin was a sound astronomer, a devoted follower of

2. See also the notes in chapters 4 and 5 regarding Peter Apian and Maestlin. Philip Apian succeeded his father, Peter, to the chair of mathematics at Ingolstadt at the age of twenty-one in 1552. Having embraced Protestantism, Philip was forced to leave the chair but was then appointed to the chair of mathematics and astronomy at Tübingen. In 1584 he was forced to resign and was succeeded by Maestlin. The Lutheran *Book of Concord* (1580) contained the nine symbolical books of the Lutheran church, including the *Formula of Concord* (1577) drafted by six Lutheran divines to end the controversies that distracted the church during the quarter century after Luther's death (1546). Its adoption led to a large secession to the Reformed Church.

Copernicus, and the observer of the cometary phenomenon. His observations cut at the underlying assumption of Aristotle's astronomy regarding the divine and unchanging nature of the heavenly sphere. It is understandable that young Kepler was filled with Copernican ideas and anti-Aristotelian physics during his student days. Even so, neither astronomy nor mathematics was his chosen vocation. In 1593, at the age of twenty-two, Kepler was still studying in the theological faculty at Tübingen when he was offered a post at the University of Gratz as a teacher of mathematics and astronomy. Apparently with some reluctance and many doubts, he accepted and arrived in Gratz to assume his new duties in April 1594. He must have been a poor lecturer because the small number of students who attended his course diminished to zero at the end of his first year. Believing that he could not retain his post, he implored Maestlin to find him another position. It is to the everlasting credit of the directors of the school at Gratz that they appreciated the brilliance of Kepler and arranged for him to give additional lectures in classical subjects in anticipation that his mathematical genius would be recognized eventually. Two years after his arrival in Gratz, he published the first of his important books, the *Mysterium cosmographicum*.[3]

In a preface to that book Kepler describes how, on July 9, 1595, when he was drawing figures on the blackboard for his small class, he was suddenly filled with the idea that the universe must be built around certain symmetrical figures. These were the five regular solids—the cube, the tetrahedron, the dodecahedron, the icosahedron, and the octahedron—all having the characteristics that the faces of each solid are equal and that the faces can be constructed from equilateral figures. This unique peculiarity of the five solids had been recognized since ancient times, but Kepler had the vision of their possible relationships to the planets. If the sphere of Saturn's orbit circumscribed the cube, if the sphere for Jupiter was placed inside this cube, the tetrahedron inside Jupiter's sphere and the

3. Tübingen, 1596.

sphere of Mars inscribed in the tetrahedron, and so on, then Kepler maintained that the relative dimensions of the spheres would be precisely those needed in the Copernican theory. Kepler's intense belief in the harmony of the world led him to conclude that he had solved the mystery of the universe. Only five perfect solids exist; that they should define the five intervals between the planets seemed to Kepler to be a divine arrangement—and, moreover, a circumstance that underlined his belief in the heliocentric nature of the universe.

The scheme of *Mysterium cosmographicum* was a false trail, but one that step by step led Kepler to his three laws of planetary motion. Although the visionary transference of the perfection of the five regular solids to the planetary system, described in this early work, had no hope of future development, nevertheless the geometrical construction immediately illuminated one of the fallacies in Copernican mathematics. Copernicus had preserved in his heliocentric scheme a fallacy of the Ptolemaic universe: that the *planes* of the orbits of the planets all intersected at the center of the Earth. Kepler's first geometrical construction, as described in *Mysterium cosmographicum,* immediately suggests that if the Sun and not the Earth is the central and governing feature of the system, then the planes of the planetary orbits must intersect at the center of the Sun. It was a most important rationalization of the Copernican scheme, which immediately accounts for the north and south deviations of the planetary orbits from the plane of the ecliptic. Further, Kepler's insistence that all orbits of planets should be computed with the Sun as both the physical and mathematical center of the system immediately eliminated the difficulty of the apparently changing eccentricities of Mercury and Venus. It was an artificial difficulty introduced by Copernicus into his heliocentric scheme because he had computed planetary eccentricities from the center of the Earth's orbit instead of from the Sun. Copernicus was forced to introduce more minor epicycles to deal with this anomaly, but Kepler showed that the problem disappeared if the Sun was taken as the reference point for the planets as well as for the Earth.

By a series of steps of this character, Kepler slowly changed the mathematical treatment of the Copernican system. Eventually, the severe problem of the orbit of Mars and Tycho's accurate series of measurements of that planet led Kepler to an immensely powerful conceptual scheme in which the ancient belief in the divine perfection of the circle had to be abandoned. The proximity of Mars to the Earth and the eccentricity of its orbit had defied all attempts to account for the changing position of the planet. It was the weakest point of the Ptolemaic theory, and the Copernican scheme did not give any substantial improvement in the predictions made on the basis of the geocentric concept. Tycho laboured long on the problem, but he could not reconcile his measurements with any scheme of planetary orbits. The inheritance by Kepler of Tycho's measurements, [4] which were accurate to four minutes of arc, was the key that eventually led him to the revolutionary concepts embodied in his first two laws of planetary motion. Twentieth-century imagination, based on the computation of planetary and space probe orbits by high-speed computers, is staggered at this immense accomplishment of Kepler—a prolonged mathematical labor that occupied him for ten years. Kepler tried endless combinations of circular motions for the orbit of Mars and the orbit of the Earth from which the planet is observed. Many of Kepler's discarded solutions were satisfactory for older observations, but they failed when applied to the new measurements of Tycho.

Finally Kepler was forced to conclude that no system or combination of circular motions could account for the motion of Mars and that some other geometrical figure must be involved. The supreme brilliance of his achievement was not only to recognize that the orbits must be elliptical—that idea alone did not solve the problem—but that the speed at which the planets moved in their elliptical orbits was governed by a simple law. In 1609, thirteen years after the publication of *Mysterium cosmographicum,* Kepler announced his results in *On*

4. Tycho and Kepler met in Benatek Castle near Prague, which Tycho had selected as his headquarters in February 1600, but they had corresponded for two years previous to the meeting.

the Motion of Mars (Astronomia nova).[5] This great work, published in Prague, enunciated the first two Keplerian laws: (1) planets move in elliptical orbits with the Sun as one of the foci of the ellipse; and (2) the orbital velocity of the planet varies so that the line joining the planet to the Sun sweeps through equal areas of the ellipse in equal intervals of time. The substitution of ellipses for circles, coupled with the simple law of motion, immediately removed all ambiguities of the heliocentric theory without the need for equants, epicycles, deferents, or other artificial geometrical constructions. Moreover, it gave predictions agreeing with observations for the first time in history. The elegance of the Keplerian scheme, based on the Copernican heliocentric hypothesis, was such that no reasonable astronomer or mathematician could any longer find either mathematical or scientific reasons for preferring the geocentric model.

Kepler's work was not finished. In 1619 he announced his third law of planetary motion in *Harmonies of the World.*[6] Whereas the first two laws concerned the orbits of individual planets, this third law defined the cohesion of the entire planetary system by relating the speeds of the planets in their differing orbits. If the times taken for two planets to complete their orbits are T_1 and T_2, and if the average distances between the planets and the Sun are R_1 and R_2 respectively, then Kepler discovered that these parameters were related so that the ratio of the squares of the orbital periods were proportional to the cubes of the average distances from the Sun—that is, $\left(\dfrac{T_1}{T_2}\right)^2 \propto \left(\dfrac{R_1}{R_2}\right)^3$. Today it seems extraordinary that Kepler's almost mystical belief in the harmonies of the heavens could lead to these revolutionary concepts. He elaborated many other mathematical relationships that were subsequently abandoned as unnecessary or inexact. For example, he deduced

5. "A new astronomy based on causation, or, A physics of the Sky derived from investigations of the motion of the star Mars founded on observations of the noble Tycho Brahe."

6. *Harmonices mundi* (Linz, 1619).

that the maximum and minimum orbital speeds of a planet were related to the concordant musical intervals. Kepler's historic work was based on two fervent beliefs: (1) that the heliocentric hypothesis was correct and (2) that God had constructed the universe in accordance with divine harmonic relationships. Accused of "throwing Christ's kingdom into confusion with his silly fancies," Kepler, filled with religious spirit, declared, "I do think the thoughts of God."[7]

7. See A. D. White, *A History of the Warfare of Science with Theology in Christendom* (London, 1955), p. 168.

7

Galileo

Galileo, born in Pisa in 1564, was a contemporary of Kepler. Remarkably, there seems to have been no contact between them until 1597. In that year Kepler sent copies of his recently published *Mysterium cosmographicum* to Galileo. In his letter to Kepler acknowledging the gift Galileo wrote:

> . . . so far I have only perused the preface of your work, but from this I gained some notion of its intent, and I indeed congratulate myself on having an associate in the study of Truth who is a friend of Truth. . . . I adopted the teaching of Copernicus many years ago, and his point of view enables me to explain many phenomena of nature which certainly remain inexplicable according to the more current hypotheses. I have written many arguments in support of him and in refutation of the opposite view—which, however, so far I have not dared to bring into the public light, frightened by the fate of Copernicus himself, our teacher, who, though he acquired immortal fame with some, is yet to an infinite multitude of others (for such is the number of fools) an object of ridicule and derision. . . .[1]

1. A. Koestler, *The Sleepwalkers* (London, 1959), pt. 4, p. 356.

Galileo was to suffer far more than ridicule and derision as the Catholic Church became aware of the increasing and unassailable evidence he produced supporting the Copernican hypothesis and revealing the errors of Aristotelian physics. The revolutionary ideas associated with Galileo were profound, and not merely because they were new, nor even because the ideas could be demonstrated by observation to be true. The difficulties arose because the new ideas were contrary both to common sense and theological teaching. Common sense tells us that we are at rest on a stationary Earth and that diurnal and seasonal effects arise because of the Sun's motion. We do not wake up at sunrise and think that the Sun is rising because we are on an Earth that is rotating us through space 17.4 miles every minute. Neither do we think as we approach midsummer that the Sun will be highest in the sky because we are being moved around it at a rate of over 1000 miles every minute in a certain orbit and at a certain inclination.

The ancient world view, conditioned by the astronomy and physics of Aristotle and Ptolemy, survived for over a milennium because it was concise and readily comprehended as a matter of daily experience. The stellar sphere carried the stars in their diurnal motion, and this sphere was the driving force for the motion of the planets. The stellar sphere defined the limits of the finite universe and its center was the absolute center of space—a position occupied by the Earth at rest. A stone dropped from the hand fell to the ground because it aspired to reach the center of the Earth *because* this was the center of the universe. Within this elemental framework the somewhat complicated motion of the planets across the sky found a reasonable explanation in the Ptolemaic scheme of deferents and epicycles.

The immense strength of the belief in the fixed Earth is demonstrated by its survival for two thousand years during which many real astronomical accomplishments occurred. For example, there was the discovery of the precession of the

equinoxes by Hipparchus in the second century B.C., and the
ingenious method for measuring the distances of the Sun and
the Moon devised by Aristarchus. The transition from the
geocentric to the heliocentric view would have been inevitable
in the light of Kepler's deduction of the laws of planetary motion. But a calm transition on a theoretical basis during which
astronomers may have suffered only ridicule and derision was
suddenly transformed by Galileo into an observational and
doctrinal conflict of great acerbity.

The climax in the transition from the geocentric to the
heliocentric view came in 1609 when Galileo first looked at the
heavens through his small telescope.[2] In *Sidereus Nuncius,* published in March 1610, Galileo described some of his observations. Today, when astronomical discoveries often make head-
lines, it is difficult to realize that only 350 years ago Galileo
contributed the first new class of qualitative data to astronomy
since ancient times. Almost instantly the additional light-
gathering power and definition of his small telescope revealed
five new features of the universe of destructive significance for
the ancient sciences. He saw the spots on the Sun and their
movement across the solar disk, the mountains and valleys of
the Moon, numberless new stars in the Milky Way, the moons
of Jupiter, and the phases of Venus. These last two observations provided an immensely powerful argument in favor of
the heliocentric theory. His discovery of the four principal
moons of Jupiter, and the rearrangement of their position on
successive nights, could have only one explanation: they were
in rapid rotation around the planet. Here in space was a
replica of the Copernican solar system, and the impact on seventeenth century imaginations was tremendous. The observation of the phases of Venus provided even more decisive technical evidence. In the epicyclic universe a terrestrial observer
should be able to see only a crescent of the planet. Of course
such an observation had never been made because to the un-

2. Two of Galileo's telescopes preserved in Florence have lenses with diameters of 4
and 4.4 cm and focal lengths of 95 and 125 cm.

aided eye the planet appeared always as a point. The complete cycle of phases that Galileo could see in his telescope provided perhaps the most decisive evidence that Venus was moving in a Sun-centered orbit.

It is a remarkable feature of history that the hundred years following 1540 held the overlapping lives of Copernicus, Tycho Brahe, Kepler, and Galileo. Remarkable because they produced the great cohesion and force required to overthrow the geocentric doctrine. The theoretical solution of the heliocentric concept by Kepler, and the simultaneous observational evidence obtained by Galileo, immediately created a system that made the survival of the geocentric doctrine impossible.

The Dynamical System of Galileo

The antagonism toward Galileo, although popularly supposed to be concerned only with his views about the motion of the Earth, was much more profound. The issue of the moving Earth had been faced more than half a century earlier, and the first two Keplerian laws of planetary motion had been derived before Galileo gave an account of his telescopic observations. Not only the hypothesis but the correct theory of planetary motion existed independently of Galileo. At the time of his observations Galileo had been lecturing for eighteen years at the University of Padua, where he had achieved fame for his discourses. His demonstrations with the telescope apparently created enormous enthusiasm. To those who were familiar with the concepts of Copernicus and Kepler the evidence of the telescope was almost superfluous. On the other hand, those who knew little of astronomy were now able to see for themselves that the universe did not conform to their common-sense beliefs. I have tried to imagine the effect of looking through a Galilean telescope for the first time in that epoch, and I find it hard to visualize such a profound intellectual experience. Perhaps the first look at the Earth from deep space, when one at last sees it as a sphere suspended in the heavens, might be a similar experience. In any case the telescope became almost a popular toy, and the fame of Galileo increased to the extent that he was given an additional salary and was received in great honor by Pope Paul V in 1611.

His tribulations lay in the future. Galileo had shattered Aristotelian doctrines over a broad front. Nowhere is this more clearly evidenced than in his work on the dynamics of moving bodies—indeed, Galileo created the science of dynamics. Imagine a projectile, an arrow shot from a bow, a stone thrown by the hand, or a missile shot from a cannon; they have all been given a single impulse. Of course we know that the projected object moves in a parabola before falling to Earth at a distance. This is inexplicable on the basis of Aristotelian physics, which holds that unless it is moved by an external push, the projectile will either remain at rest or move in a straight line toward the center of the Earth.

It was obvious to Aristotle that projectiles did not move in a straight line toward the center of the Earth when their contact with the initial projector was broken. He therefore argued that the horizontal motion arose because in some manner the disturbed air was the source of the push that prolonged a projectile's flight. When this was exhausted, the projectile would fall suddenly to the ground. Medieval commentators on the physics of Aristotle had already questioned this explanation, but it was Galileo who first enunciated the concept of the independence of vertical and horizontal motions. The motion of the projectile is compounded of two separate independent effects. Galileo showed that, but for the resistance of the air, the projectile would move with a constant horizontal velocity, and that the vertical fall is added independently with a velocity that increases according to the law of falling bodies. Galileo compounded these two motions to show that the path of the projected body would be a parabola and that the range in the horizontal plane would be greatest if the angle of projection was 45 degrees.

This Galilean concept—that any body will continue to

move in a straight line with uniform velocity unless it is subjected to some external force—was eventually embraced as the
principle of inertia in Newton's first law. Galileo's statement
of the relationship between the change in motion and the applied force found its quantitative expression in Newton's second law. This dynamical principle suppressed a major argument against the heliocentric hypothesis. According to its opponents, if the Earth is rotating, then a stone dropped from
the hand or from a high tower should not fall vertically but to
the west of the vertical, because the Earth has rotated during
the time of fall. This is contrary to experience. The composition of independent motions, enunciated by Galileo, accounts
for this simply because the stone retains the velocity of rotation of the Earth it possessed before it was released from the
hand.

Galileo had been firmly educated in the scholastic and
Aristotelian traditions, but by the age of twenty-one he had
already commenced a systematic investigation of Aristotle's
mechanical doctrines. By the year 1590, almost twenty years
before his telescopic observations of the heavens, he had accumulated many records of experiments on falling bodies. According to the physics of Aristotle, heavy bodies fell to the
ground faster than light bodies. For two millennia this seemed
to be a doctrine in strict accordance with observation. For us,
too, it is a matter of common observation that if we lean over
a cliff and release a stone and a feather at the same instant, the
heavy stone will reach the sea long before the feather does.
Perhaps one of the most remarkable of Galileo's accomplishments was his recognition of the superficiality of this observation. His own records of freely falling bodies, combined with a
chain of logical arguments, convinced him that whatever the
nature of a falling body—whether it be a lump of lead or a
bird's feather—the acceleration would be the same and that
any observed differences arise because of the differential effects of the resistance of the air on the body. Although
Galileo's name is forever linked with the demonstration using
the Leaning Tower of Pisa in 1591, the evidence that he car-

ried out this experiment is uncertain. It would be far easier for
an opponent to drop bodies of unequal weight from the tower
and prove that the heavier reached the ground first, thus supporting the Aristotelian view. Unless one has some means of
creating at least a partial vacuum, a simple demonstration of
the correctness of the Galilean law is difficult. Galileo argued
that in a vacuum, where the effects of air resistance were
absent, his law would be seen to be correct. When his pupil
Torricelli succeeded in producing a vacuum after Galileo's
death, the straightforward demonstration of the law of freely
falling bodies became possible. We may observe that not only
the enunciation of the law that freely falling bodies suffer the
same acceleration but also the implication of the existence of
a vacuum were contrary to the teachings of Aristotle and to
everyday experience.

The scope of Galileo's scientific work is extraordinary. In
1583, at the age of nineteen, he noticed that the great bronze
lamp hanging from the roof of the cathedral at Pisa seemed to
take the same time from one extremity of its swing to the other, whatever the extent of the swing. This soon led him to the
important discovery of the isochronism of the pendulum.
Many consider that his work on the dynamics of falling bodies
and projectiles exceed in importance his studies of the
heavens. Most of this work was accomplished before he was
appointed to a professorship at the University of Padua when
still only twenty-eight, but sometimes his views were erroneous. For example, he was wrong about the cause of ocean
tides[3] and about comets. His attitude about comets seems to
have been determined by his annoyance with a Jesuit, Father
Grassi, who had expressed the correct view that comets move
in orbits, like planets. Galileo proceeded to argue that comets

3. In his *Dialogue of the Two Principal Systems of the World* Galileo makes Salviati argue
on the fourth day that the tides were analogous to the movement of the drinking water
in the barges on the Venetian Lagoon (i.e., it flowed backward or forward when
barges were accelerated or retarded). By analogy he argued that ocean tides resulted
from a combination of the two motions of the Earth—around the Sun and about its
own axis. Kepler had already correctly concluded that the major tidal effects were
related to the Moon.

were not real objects but optical illusions. He also evolved an incorrect theory about the bending of beams under stress. He was correct in his dynamical and astronomical ideas, and these vital conclusions shattered the foundations of the entrenched Aristotelian doctrines about the world.

The Intellectual Dilemma

In some ways it is strange that the dynamical and astronomical work of Galileo had such a cataclysmic effect on the intellectual and theological outlook of the seventeenth-century world. It was not the first time that the validity of Aristotle's physics had been questioned. In his study of the Copernican revolution, Thomas Kuhn[4] established the striking similarity of the arguments used by Galileo in his *Dialogue of the Two Principal Systems of the World* and those used in the fourteenth century by Nicole Oresme. Oresme's teacher, Jean Buridan, had already demolished Aristotle's arguments about the motion of projectiles and had stated the contrary view that "the movement of the stone continuously becomes slower until the impetus is so diminished or corrupted that the gravity of the stone wins out over it and moves the stone down to its natural place"—a standpoint enunciated more mathematically by Galileo three hundred years later. As far as the motion of the Earth is concerned, although it is true that Copernicus suffered bitter comments from the Protestants, neither he nor Kepler caused the disruption associated with Galileo, although their work clearly established the heliocentric concept independently of Galileo's observations.

There were three other important features of the Galilean impact. First, whereas Oresme maintained that no argument could disprove the possibility of the Earth's motion, but that the choice between a fixed or rotating Earth must be a matter

4. T. S. Kuhn, *The Copernican Revolution* (Cambridge, Mass: Harvard University Press, 1957). See also chap. 3.

of faith, Galileo allowed no such option. People could see the
evidence for themselves, particularly in the phases of Venus
and the obvious parallelism with the moons of Jupiter. Second, the Galilean attack ranged over the entirety of Aristotelian dogma; moreover, his dynamics enabled a coherent explanation of the heliocentric system to be produced. But perhaps the most significant point is that Galileo had achieved
widespread public fame. Even before his telescopic discoveries, he commanded immense audiences. In his eighteen
years at Padua he was so successful that his salary was successively raised from 180 to 1000 florins. His lectures attracted
people of the highest distinction from all over Europe, and
eventually his university lectures had to be transferred to a
hall that could accommodate 2000 people. It is doubtful
whether any academic has had such a consistently large public lecture audience until the advent of broadcasting and television in this century. Long before his telescopic observations,
Galileo was a renowned figure, and to this fame in 1610 he
added his sensational discoveries with the telescope. Suddenly
everything seemed to conspire and cohere against the
Aristotelians. Scholars and astronomers were increasingly
forced to recognize the errors of the Aristotelian dialectic, and
a larger public than ever before in the history of Europe became interested in astronomy, and this at a time when the
mechanical nature of the universe seemed demonstrable to
anyone who cared to look at the heavens through a small telescope.

The Theological Dilemma

The recovery of the works of Aristotle in the twelth and
thirteenth centuries was a critical factor in the resurgence of
learning in that epoch. The new universities, especially in Paris, like the religious orders that had their genesis at that time,
were essentially concerned with theological problems, and the
Aristotelian outlook on the world was incompatible with
much of the established teaching. The medieval world owed

the resolution of these conflicts to the genius of Thomas Aquinas.[5]

The problems confronting the world of the seventeenth century when Galileo unleashed his attack on the entire physics and astronomy of Aristotle arose fundamentally because of the deep compatibility Aquinas had established between science and theology. The integration was such that any attack on Aristotle's science was inevitably seen as an attack on the dogma of the Catholic Church. There had been enough warnings in previous centuries of the reaction to those who produced critical commentaries on Aristotle's science. Religious dogma was especially sensitive to suggestions that the Earth was not fixed at the center of the planets and stars, or that the universe was not finite in time and space, or that heavenly bodies did not conform either in motion or shape to the perfection of the circle and the sphere. In the fifteenth century Nicholas of Cusa elaborated the idea that the universe was an infinite sphere, but he was at pains to emphasize that the argument arose because no smaller sphere would be consistent with the creative omnipotence of God. But when Bruno, in 1584, interpreted Cusa's views in terms of the Copernican hypothesis and maintained that the Sun, Earth, and planets need not be the center of an infinite universe, or indeed may not be unique, he was held to be a pagan, imprisoned by the Inquisition, and finally burned at the stake in 1600. Galileo's telescope revealed the seeming infinite extent of the stars, that the Earth was not fixed, that heavenly motions were not circular, and that heavenly bodies were not perfect—the Sun had spots and the Moon was defaced with mountains and valleys.

The Breaking Point

As the second decade of the seventeenth century opened, we see the absolute conflict with established dogma. No longer

5. See chap. 2.

were issues localized to erudite discussions of scholars on the
basis of hypothetical alternatives, for the overriding authority
of the Catholic Church was in jeopardy. It was inevitable that
either Galileo and the new science and astronomy would have
to be dishonored and suppressed or there would have to be a
reinterpretation of the Scriptures to embrace the new concepts
following the precedent of Aquinas, four centuries earlier.

Galileo was denounced, and the teaching of the heliocentric
doctrine was prohibited in 1616. In the famous trial of 1633
the Council of the Holy Office condemned Galileo to prison
and ordered him to recant. These events happened, but why
they occurred is not clear. The Catholic Church had already
displayed reasonable wisdom over its attitude to the heliocentric hypothesis; *De revolutionibus* was freely available and its
principles taught in some major Catholic universities. The reformation of the calendar by Gregory XIII in 1582 used data
based on the Copernican system, and there had been no suggestion that the Church should impose cosmological conformity. Even the mind of the aged chief theologian of the
Church, Cardinal Bellarmine, was not closed. In 1615 Father
Foscarini, a Carmelite monk of excellent reputation, sent a
document to Bellarmine in which he suggested that it was
time for the heliocentric doctrine to be accepted as a physical
reality and suggested measures by which the relevant passages
of Scripture could be reconciled. Bellarmine's long reply to
Foscarini on April 12, 1615, is a revealing document. He re-
minds Foscarini that the man who wrote "The Earth abideth
for ever; the Sun also riseth, and the Sun goeth down, and
hasteth to his place whence he arose" was Solomon, who
spoke by divine inspiration and acquired all this wisdom from
God Himself. Nevertheless, Bellarmine writes, "If there were
a real proof that the Sun is in the centre of the universe, that
the Earth is in the third Heaven, and that the Sun does not go
round the Earth but the Earth round the Sun, then we should
have to proceed with great circumspection in explaining passages of Scripture which appear to teach the contrary, and
rather admit that we did not understand them than declare an

opinion to be false which is proved to be true."[6]

But within months the apparatus of the Inquisition was unleashed upon the heliocentric doctrine and Galileo. Scholars
of the twentieth century who have gained access to the
archives of this epoch remain puzzled as to why the conflict
occurred. A major faction of church intellectuals were on
Galileo's side, and Bellarmine was certainly not inflexible.
The conclusion of De Santillana is that "the tragedy was the
result of a plot of which the hierarchies themselves turned out
to be the victims no less than Galileo—an intrigue engineered
by a group of obscure and disparate characters in strange col-
lusion who planted false documents in the file, who later mis-
informed the Pope and then presented to him a misleading
account of the trial for decision."[7] Galileo had many powerful
enemies, and no doubt he himself behaved in an arrogant
manner after the first injunctions of 1616. Doubtless, too,
many other pressures were bearing on the Church as a result
of the Protestant revolt. In any event, the final judgment delivered on behalf of Urban VIII, on Midsummer Day in 1633,
can be seen as a lasting tragedy for the civilized world. Not
until 1822 did the Church remove the ban on the printing of
books that treated the Earth's motion as physically real. To
appreciate the absurd position into which the Church had
been forced by Urban VIII, it must be remembered that the
official commitment to the belief in a fixed Earth still existed
when the Royal Astronomical Society was formed, and for 162
years after the foundation of the Royal Society. Irrevocable
harm was done to the prestige of the Church and the conflict over human understanding and purpose remains unresolved.

Galileo was born in 1564, the year of Michelangelo's death.
He died in 1642, the year of Newton's birth. His life span of
seventy-eight years marks a turning point in history from the
epoch when man's comprehension of the universe could be

6. The full text of this letter is given by Giorgio de Santillana, *The Crime of Galileo*
(Chicago, 1955; London, 1958), pp. 98–100.

7. Ibid., pp. xii–xiii.

ral circular motions. His solution to the question why planets move in an ellipse with the Sun at one of the foci was ingeniously based on the priority he attached to the Sun "who alone appears, by virtue of his dignity and power, suited for this motive duty and worthy to become the home of God himself, not to say the first mover."[1] He visualized solar forces as an *anima motrix*—a system of rays emanating from the Sun and localized to the plane of ecliptic. These rays rotated with the Sun, and by pushing against a planet they would impel and maintain the planet in circular motion. Nevertheless, his laws of planetary motion required that the planets did not move in circles but in ellipses. Kepler ingeniously invoked a second force that would alter the distance between the planet and the Sun at different parts of the orbit. At about that time, in 1600, William Gilbert had published *On the Magnet,* in which he recognized that the Earth was a huge magnet. Kepler generalized this idea to include the other planets of the solar system and proposed that this magnetic attraction and repulsion between the Sun and the planets was the second force that changed the circular motion engendered by the *anima motrix* into elliptical orbits. As he also envisaged that the *anima motrix* was strongest near the Sun and decreased as the distance from the Sun increased (so that the planet's orbital speed was inversely proportional to its distance from the Sun), he envisaged in a qualitative manner the law of force that was to be defined precisely by Newton eighty years later.[2] Between Kepler and Newton, three men—Descartes, Borelli, and Hooke—exerted powerful influences on the development of ideas about such forces and motions.

1. Quoted by T. S. Kuhn, *The Copernican Revolution* (Cambridge, Mass: Harvard University Press, 1957), p. 130, from a fragment of one of Kepler's early disputations trans. E. A. Burtt, *The Metaphysical Foundations of Modern Physical Science* [New York, 1932].

2. In his idea that magnetic forces between the Sun and the planets were a controlling influence in the solar system, Kepler also presaged certain theoretical developments of recent years in which magnetic torques in the solar system have been invoked to explain the anomaly that the rotational momentum of the solar system is concentrated in the planets and not in the Sun, which is 750 times more massive than the whole of the remaining planetary system.

René Descartes, who lived from 1596 to 1650, occupies a distinguished position in the development of philosophical thought, in mathematics, and in science. His *Principia philosophiae,* published in 1644, contains the essence of the ideas of motion in a Copernican universe. For the first time he considers how a single corpuscle moves in a void and inquires how the motion might be changed by collision with another corpuscle. He arrives at two extremely important conclusions: (1) a corpuscle in a void will remain at rest, and (2) a corpuscle in motion will continue in motion at the same speed and in a straight line unless deflected by another corpuscle. These are the first statements in history of the law of inertial motion —and, of course, they are entirely antithetical to the concept of a self-sustaining circular motion envisaged by Copernicus.

Descartes developed this contemporary version of the ancient idea that the motions of particles were governed by laws imposed by God at the Creation. Descartes was the first to enunciate in a decisive manner the concept that for a planet to follow a closed path around the Sun, it must in some way continually fall toward the Sun so that its inertial linear motion is transformed into motion in a curve. Descartes did not solve the problem of the nature of the force that caused this continued fall. He believed that the planets were pushed toward the Sun by corpuscular impact, and his theory that the primeval corpuscles would eventually suffer collisions so that they formed into vortices filling the entire universe and became potential planetary systems was not a fruitful extension of his law of inertial motion.

In his philosophical works, especially his *Discourse on Method* (1637) and *Meditations* (1642), Descartes laid many of the foundations for subsequent philosophical development. His thought dominated much of the outlook of the seventeenth century not only in philosophy but also in science and cosmology. He developed metaphysical and other proofs for the existence of God, maintaining that God, as the most perfect Being, is better if He exists than if He does not, and hence if God does not exist, He is not the most perfect Being. Notwithstanding his arguments for the existence of God as the Creator of matter, and that the act of Creation was the same as that by which creation is sustained, Descartes was viciously attacked by the Protestants. In particular the theologians of Holland attempted to have him tortured to death on a charge of atheism. The Catholic theologians of France were also bitterly opposed to him. Even before Descartes's publications unleashed these attacks, he was well aware of the danger of expressing conformity with the heliocentric concept. His cosmological ideas were ready for publication in 1633, but he withdrew his book when he heard of the trial and judgment of Galileo. His *Principia philosophiae* did not appear for another eleven years, by which time both the *Discourse* and the *Meditations* had been published.

The widespread accomplishments of Descartes were paralleled in other spheres by his Italian contemporary G. A. Borelli, who lived from 1608 to 1679. Borelli was a friend and disciple of Galileo whose scientific activities ranged over the fields of physiology, mathematics, and astronomy.[3] As far as the motion of the planets was concerned, Borelli did not believe that any "push" of the kind envisaged by Kepler for the *anima motrix* could keep a planet moving in a closed orbit around the Sun. He argued that some force must exist to pull the planet toward the Sun, otherwise the planet would move off at a tangent and leave the solar system. Historically this conception of Borelli's was the first suggestion that there was some force acting at a distance.

Robert Hooke was the youngest of the three who prepared the way for the great synthesis of Newton. There is some dispute whether Hooke or Newton first arrived at the idea of universal gravitation, namely, that the force that controls bodies on Earth is the same as the force governing celestial motions. Hooke was born on the Isle of Wight in 1635, seven years before the birth of Newton and the death of Galileo. He had no material advantages, and he was poor, weak, and unattrac-

3. The aspect of physiology that treats of muscular movements on mechanical principles was founded and developed by Borelli.

tive physically. When he was eighteen years old he entered the University of Oxford and came under the influence of Robert Boyle, who employed Hooke as a paid assistant. Immediately Hooke excelled in the new activity of experimental science; he devised and constructed the first air pump to be made in England, he invented a watch with a balance wheel controlled by a spring, and there is good evidence that the discovery of Boyle's Law should be attributed to Hooke.[4] In 1662, when the Royal Society received its charter from Charles II, Hooke was made curator. He lived at Gresham College, where he was appointed professor of geometry in 1665, and from there he carried out his astronomical work. In that year he published his *Micrographia*[5] displaying an amazing sweep of activity in many aspects of the biological and physical sciences. Here we are concerned only with his astronomical work. Observationally he developed and improved many astronomical instruments including the first reflecting telescope of the Gregorian type.[6] He discovered that Mars and Jupiter were rotating, recognized the existence of double stars, realized that the tail of a comet was always repelled by the Sun, and in attempting to measure stellar parallax found anomalies later explained when Bradley discovered the phenomenon of aberration.[7]

In 1666 Hooke gave a significant practical demonstration to the Royal Society. He suspended a heavy pendulum bob by a

4. See E. N. Da. C. Andrade, "Robert Hooke FRS," *Notes and Records of the Royal Society* 15, (1960): 137.

5. *Micrographia or Some Physiological Descriptions of Minute Bodies made by Magnifying Glasses with Observations and Inquiries thereupon* (London: Royal Society, 1665).

6. The Gregorian telescope was designed by the Scottish mathematician James Gregory (1638–1675), but Hooke appears to have made the first practical instrument. In this telescope the light is reflected from the primary parabolic mirror to a concave (ellipsoidal) secondary mirror. From this secondary mirror the light is reflected back, along the axis, through a central hole in the primary mirror to the eyepiece or recording system mounted behind the primary. The more frequently used Cassegrain system is a similar arrangement, but in this case the secondary mirror is convex (hyperboloidal). The great advantage of these systems is that with long-focus telescopes, where access to the primary focus is difficult, the observer or other apparatus can be stationed near the base of the telescope.

7. See chap. 11.

wire so that it was free to move in any direction. When he pulled the bob from its position of rest and released it, it simply oscillated to and fro in a plane, like an ordinary pendulum. When he gave the bob a horizontal push, however, it executed a continuous orbit about its position of rest (the conical pendulum), precisely like the orbit of a planet around the Sun. Thus Hooke gave a clear demonstration that the forces at work were unable to pull the bob to the center but always produced a motion so that it moved in a curve around the central point. A similar single centrally directed force in the heavens, argued Hooke, would have precisely the same effect in giving rise to the orbital motion of the Earth and planets around the Sun.

In 1674 Hooke published his theory that there was a force of universal gravitation and maintained that the planetary system was one "answering in all things to the common Rules of Mechanical Motions."[8] The idea of a force keeping the planets in orbit, invoked by Kepler, Descartes, and Borelli, had involved in all cases some element of celestial mystery. Hooke was the first to postulate that there was no more mystery in the heavens than on Earth about such forces. He stated his view that the planets moved in closed orbits around the Sun because they were subject to a gravitational pull and that the strength of this force decreased with distance.

> All celestial bodies whatsoever have an attraction or gravitating power towards their own centres, whereby they attract not only their own parts, and keep them from flying from them, as we may observe the earth to do, but that they do also attract all the other celestial bodies that are within the sphere of their activity; . . . all bodies whatsoever that are put into a direct and simple motion, will so continue to

8. *An attempt to Prove the Motion of the Earth from Observations* (London, 1674) included as the first of the series of papers published as *Lectiones Cutlerianae or a Collection of Lectures Physical Mechanical Geographical and Astronomical Made before the Royal Society on several occasions at Gresham College. To which are added divers Miscellaneous Discourses* (London, 1679). These *Lectiones Cutlerianae* were named after Sir John Cutler, who promised money to Hooke that was never paid.

move forward in a straight line, till they are by some effec-
tual powers deflected and bent into a motion, describing a
circle, ellipse, or some other more compounded line curve
. . . these attractive powers are so much the more powerful
in operating, by how much nearer the body wrought upon
is to their own centres. Now what these several degrees are
I have not yet experimentally verified. . . .

Hooke thus gave a clear enunciation of the force of universal
gravitation, but he was unable to establish the "degrees," or
precise law, of this force and relate this to the observed elliptical orbits of planets around the Sun. This final synthesis was
the immense achievement of Newton.

Hooke and Newton were contemporaries, the former being
seven years older, and not only their lives but also their scientific activities were intertwined. The overlap was a cause of
bitter dispute between them on at least two separate issues—
concerning optics and gravitation. Newton was led to the idea
of using a mirror instead of a refracting lens in a telescope
because of the chromatic aberration he had observed in his experiments with lenses. In an instrument he exhibited before
the Royal Society in 1671, the small concave mirror was
ground from speculum metal.[9] The light was reflected back
along the axis of the tube; at the focus of this primary mirror
Newton placed a plane mirror at 45 degrees so that the image
was deflected at right angles to the axis of the tube and could
be readily viewed by an observer through an eyepiece at the
periphery of the tube.[10] Newton claimed that this telescope,
which was only six inches long, magnified forty times. He
knew about Gregory's design for a reflecting telescope, but he
probably did not know that Hooke had constructed a
Gregorian telescope. Hooke was stimulated to claim priority
for the reflecting telescope he had made in 1664 "but the
plague happening which caused my absence and the fire . . . I

9. An alloy of tin, arsenic, and copper that Newton made himself.
10. Compare the reflecting arrangements for the Gregorian and Cassegrain telescopes on page 93, fn. 6.

neglected to prosecute the same, being unwilling the glass grinders should know anything of the secret."

A year later, in 1672, when Newton announced to the Royal Society his theory that white light was a mixture of many colors, Hooke again opposed him, saying that he too had made such experiments and that Newton's explanation was not correct. It seems that Hooke's opposition to Newton's optical work was the primary cause of the delay in publication of Newton's great work *Opticks*. The work was not published until 1704, a year after the death of Hooke, although it had been in manuscript for many years. By the time of publication, Newton had long ceased to investigate the subject.

An immense amount of modern scholarship has been devoted to the events leading to the development and publication in 1687 of *Principia mathematica*.[11] Newton graduated from the University of Cambridge in 1665, the year in which the Great Plague struck the country.[12] In the autumn the university was closed, and Newton returned to his birthplace at Woolsthorpe in Lincolnshire. He remained there in seclusion for nearly two years while he carried out his work in optics and calculus, but above all it was the period when he laid down the foundations of the theory of gravitation. Toward the end of his life he wrote: "In the same year [1666] I began to think of gravity extending to the orb of the Moon . . . and deduced that the forces which keep the Planets in their Orbs must [vary] reciprocally as the squares of their distances from the centres about which they revolve. . . . All this was in the two plague years of 1665 and 1666, for in those days I was in the prime of my age for invention, and minded mathematics and philosophy more than at any time since."[13]

The procedures by which Newton was led to this law have

11. *Philosophiae naturalis principia mathematica.*
12. No effective medical treatment was known. In London, more than a tenth of the population died in three months.
13. For the text and commentary on this memorandum, see I. B. Cohen, *Introduction to Newton's Principia* (Cambridge, England: Cambridge University Press, 1971), pp. 291–92.

been discussed by I. B. Cohen.[14] Newton knew of Descartes's law of inertia: in the absence of an external force a body at rest will continue to be at rest, while a body moving uniformly along a straight line will continue to do so. He had learned of Kepler's laws by reading Streete's *Astronomie Carolina*. And as Hooke had done, Newton realized that the law of inertia and the nature of the planetary orbits required a Sun-centered force for the motion of planets and a planet-centered force for the motion of satellites. In the *Principia* Newton restates Descartes's law of inertia, without acknowledging Descartes's priority, as the first law of motion. In the "definitions" with which the *Principia* opens Newton states two further laws of motion. Law II states that the change in motion of a body is proportional to the force acting on it, and Law III is the law of equal action and reaction. Newton claims that he was led to these laws by studying the works of Galileo, Wallis,[15] Wren, and Huyghens.

The major elements of Newton's laws of motion and gravitation were already at hand, and in 1679 or 1680 he recognized the vital fact that his inverse square law of force, which he had discovered in 1666 as the form of the force acting between two bodies, led mathematically to the Keplerian laws of planetary motion. It is not clear why, having discovered the nature of the inverse square law in 1666, Newton did not immediately recognize its significance as the explanation of the Keplerian laws of planetary motion. Some analysts believe that the figures Newton used for the Moon's distance were so inaccurate that he could not make the theory fit the observations. It seems more likely that, although he knew that every particle attracted every other particle according to the inverse square law, he was uncertain how this idea should be applied when the attraction of large bodies like the Earth and the other planets were involved. The problem of adding all the forces of attraction for every particle in the large body defied Newton

14. *Notes and Records of the Royal Society* 19 (1964): 131.
15. John Wallis, mathematician (1616–1703), is credited with the introduction of the infinity symbol.

for several years until he eventually discovered the solution, namely, that the attraction of a sphere for a particle is the same as if the entire mass of the sphere were concentrated in a single point at its center.

Today we regard this as obvious, but in 1685 it appeared as a surprising conclusion. More than twelve years had elapsed between Newton's discovery of the inverse square law of gravitation and his realization that, by treating the planets mathematically, as though their entire mass were concentrated at the center of the planet, the laws of planetary motion were an inevitable consequence. Newton had not yet published his discoveries and might never have done so but for the persistence of Hooke and Halley.

In the interval between 1666, when Newton discovered the inverse square law of gravitation, and 1685, when he realized how to apply the law to planetary orbits, there was a somewhat unfriendly correspondence between Hooke and Newton. It seems clear that neither realized their mutual interest in this issue, and in ignorance of the fact that Newton had arrived at the solution of the inverse square law in 1666, Hooke proposed in his correspondence several years later that the law of force between two bodies might be of this type. As far as Edmond Halley was concerned, his interest in the subject of planetary orbits does not seem to have emerged in any practical manner until 1684. He was then twenty-eight years old and he had already been engaged on his voyages to observe the transit of Mercury and Venus, had made his famous pronouncement about the orbit of the great comet of 1680 (subsequently known as Halley's comet), had published papers on terrestrial magnetism, and had begun to study the phenomenon of tides.[16] In 1684 Halley's interest turned to the problem posed by Kepler's third law: why the square of the time for a planet to orbit the Sun was related to the cube of its average distance from the Sun. Halley concluded that if the Sun attracted each planet by a force that

16. An account of Halley's remarkable career is given by Colin A. Ronan in *Edmond Halley, Genius in Eclipse* (New York, 1969; London, 1970). Halley died in 1742. In 1720, when he was sixty-four, Halley was appointed the second Astronomer Royal.

varied inversely as the square of the distance between them, then precisely this relationship would result. He discussed this problem with Hooke, who claimed to have reached the same conclusion. But Hooke offered no proof, and so Halley decided to visit Newton at Trinity College.

The historic meeting of Halley and Newton occurred in August 1684. Newton agreed that the inverse square law was correct, but he could not find the papers and promised to send them to Halley in London. John Conduitt (husband of Newton's niece) recorded the climax of this meeting, as Halley asked Newton what the orbit described by a planet would be if gravity diminished as the inverse square of the distance.

In 1684 Dr. Halley came to visit him at Cambridge, after they had been some time together, the Dr. asked him what he thought the Curve would be that would be described by the Planets supposing the force of attraction towards the Sun to be reciprocal to the square of their distance from it. Sr. Isaac replied immediately that it would be an Ellipsis, the Doctor struck with joy and amazement asked him how he knew it, Why saith he I have calculated it, Whereupon Dr. Halley asked him for his calculation without any further delay, Sr. Isaac looked among his papers but could not find it, but he promised him to renew it and then to send it him. . . . [17]

When he did so, Halley, realizing the immense significance of Newton's idea, returned again to Cambridge. It was then that Newton promised to publish his work. At the meeting of the Royal Society on December 10, 1684, Halley gave a report on Newton's work, and in the *Philosophical Transactions* of 1686 Newton's paper on the theory of gravitation was published. In May of that year the Council of the society ordered the book on which Newton was working—the *Principia*—to be published. Difficulties arose from a series of controversies with

17. For the context and source of this extract, see Cohen, *Introduction to Newton's Principia*, p. 50.

Hooke, and from the fact that, with Halley's encouragement, the *Principia* was emerging as a far more substantial work than originally envisaged. In 1687, when parts of the *Principia* were already at the printer's, the Royal Society decided it could not meet the costs of publication. At this point, Halley, certain of the importance of Newton's work, offered to check the proofs and pay the printer himself.[18] The dispute with Hooke about whether he or Newton claimed priority for the discovery of the inverse square law became so serious that Newton threatened that he would not write the third book of the *Principia* unless the Royal Society agreed not to recognize Hooke's claims. The intervention of Halley as the person through whom most of these disputes were conducted eventually smoothed Newton's irritation to such an extent that the *Principia* was completed, but without acknowledgment to Hooke. In July 1687, the *Principia* was published by the Royal Society.[19] The book that Halley referred to as the "divine treatise" seemed then, as today, one of the great scientific texts of all time—an immense and universal synthesis linking terrestrial events with the motion of the Earth and all celestial bodies.

Two years after the publication of the *Principia* Newton was elected to Parliament as member for the University of Cambridge. He was made Master of the Mint in 1699, and became president of the Royal Society in 1703. He died in 1727, a year after the publication of the third edition of *Principia*. Throughout these latter stages of his career Newton continued to be involved in various controversies, especially with Leibniz over the question of the invention of calculus. During Newton's life few mathematicians understood calculus and few scientists had mastered the contents of the *Principia*. French mathematicians were the first to exploit and extend the principles enunciated by Newton, especially Joseph La-

18. A truly incredible state of affairs. The Royal Society was in such poor condition financially that it could not pay Halley's salary of 50 pounds per annum in cash. The minutes of the council record that Halley was to be given fifty copies of Willoughby's *Historia piscium* instead of the 50 pounds.
19. Unbound copies of the book sold for six shillings. An original edition changed hands in 1979 at 5000 pounds sterling.

grange (1736–1813), Pierre Laplace (1749–1827), and Jean d'Alembert (1717–83). To these names should be added the Swiss mathematician Leonhard Euler (1707–83), who explained the irregularities in the Earth's movement since the time of Ptolemy.[20] The mechanistic concept of the universe based on Newtonian theory as developed by these mathematicians and astronomers during the eighteenth century was exemplified in the work of Laplace. His thirteen volumes of *Mécanique céleste*[21] provided a demonstration that the universe was a huge machine governed by the Newtonian laws. It was not a development that would have pleased Newton. A century earlier, in the *Principia,* he discussed the nature of God and concluded that we know Him "by His most wise and excellent contrivances of things, and final causes."

20. The axis of the Earth's orbit had altered by 5 degrees since that time. Euler went blind early in his career.
21. *Traité de mécanique céleste* (1799–1825).

9

William Herschel and the
Emergence of Modern Cosmology

The revolutions in human thought associated with Copernicus
and Galileo in the sixteenth and seventeenth centuries did not
immediately introduce any substantial change in man's wider
concepts about the nature of the universe. In the cosmic sense
these revolutions were local. In Aristotle's cosmology the
Earth was stationary at the center of the universe, and matter
and space ended at the sphere of the fixed stars. This ancient
cosmology was both geocentric and egocentric. After the age
of Copernicus and Galileo, cosmology became heliocentric but
remained substantially egocentric. That is, the Sun, instead of
the Earth, had to be regarded as the center of the universe; but
in the overall cosmological outlook, the universe consisted of
the sphere of the fixed stars with the Earth, in motion around
the Sun, in the central region.

Although the heliocentric universe remained essentially
egocentric, there were consequential changes in outlook of far
greater significance than that of the local motion of the Earth
around the Sun. In the ancient cosmology the sphere of the
stars defined an absolute frame of reference. Objects fell toward the Earth because it was the fixed center of this sphere,
whereas the sphere of the fixed stars provided the motive force
for the movement of planets. In the Copernican universe these

explanations became untenable. Objects still fell to the Earth, but the Earth was in motion and could no longer be regarded as the absolute center of the universe. Further, constraints concerning the necessity for the sphere of the fixed stars to be in motion and finite were removed.

The removal of these constraints had a profound influence on the evolution of ideas about the nature of the universe. The belief in an infinite universe emerged logically when the heliocentric theory removed the constraints requiring that a sphere of fixed stars surrounded the Earth. In fact, Aristotle's proof that there was no void outside the stellar sphere had never been generally accepted; and long before the age of Copernicus, the popular view seems to have been in favor of an infinite universe beyond the sphere of the fixed stars—an infinite space containing no matter but serving as the abode of God and the angels.

Soon after the publication of *De revolutionibus,* the first steps emerged in the integration of the concept of the infinite universe with the heliocentric theory. In 1576 a unique work by Thomas Digges appeared. His father, Leonard, was the author of the popular almanac *A Prognostication everlastinge,*[1] and in the edition of 1576 a supplement by Thomas appears entitled *A Perfit Description of the Caelestiall Orbes according to the most aunciente doctrine of the Pythagoreans, lately revived by Copernicus and by Geometricall Demonstrations approved.* The significance of this work was that it contained the first diagram of the universe in which the stars are no longer shown as constrained to a fixed sphere. The legend on the diagram states: "This orbe of starres fixed infinitely up extendeth hitself in altitude sphericallye . . ." and continues with the concept that the stars are like the Sun ". . . perpetuall shininge glorios lightes innumerable. Farr excellinge our sonne both in quantitye and qualitye the very court of coelestiall angelles. . . ."

This idea that stars were scattered through infinite space received substantial observational support when Galileo dis-

1. London, 1576. Includes the supplement by his son.

covered countless new stars with his telescope where none could be seen with the unaided eye. The ideas of inertia and gravity introduced by Descartes and Newton gave scientific validity to this concept of an infinite universe. Indeed, as evidence accumulated that gravitational laws were applicable to the stellar system and not merely to the local system of planets, belief in the infinite extent of the universe became a necessity in order to preserve its stability. Otherwise a purely local system of stars would have contracted toward their center of mass under the self-gravitational forces. Newton was certainly clear on this point, as his correspondence with Bentley makes plain.

Richard Bentley was a brilliant scholar who in 1692, at the age of thirty, gave the first of the lectures in memory of Robert Boyle.[2] Bentley delivered the lecture *"A Confutation of Atheism"* at St. Martin-in-the-Fields, and it seems probable that this was the first occasion on which Newton became known to the general public. In any event, Bentley entered into an extensive correspondence with Newton because the theme of the lecture was that Divine Providence was evident to any who could perceive the design of the universe as revealed by Newton. On December 10, 1682, Newton wrote Bentley about the infinite extent of the universe.[3]

> But if the matter was eavenly diffused through an infinite space, it would never convene into one mass but some of it convene into one mass & some into another so as to make an infinite number of great masses scattered at great distances from one to another throughout all yt infinite space. And thus might ye Sun and Fixt stars be formed supposing the matter were of a lucid nature. But how the matter

2. Boyle died in 1691. In his will he established an annual lecture "for the defence of religion against infidels."
3. H. W. Turnbull, ed., *The Correspondence of Isaac Newton* (Cambridge, England: Cambridge University Press for the Royal Society, 1961), vol. 3, letter 398. This passage is contained in the first of four letters written to Richard Bentley at a time when he was prebendary and chaplain to the Bishop of Worcester. The original manuscript is in Trinity College Library, Cambridge.

should divide it self into two sorts & that part of it wch is fit
to compose a shining body should fall down into one mass
& make a Sun & the rest wch is fit to compose an opake
body should coalesce not into one great body like ye shining
matter but into many little ones: or if the Sun was at first an
opake body like ye Planets, or ye Planets lucid bodies like ye
Sun, how he alone should be changed into a shining body
whilst all they continue opake or all they be changed into
opake ones whilst he remains unchanged, I do not think
explicable by mere natural causes but am forced to ascribe
it to ye counsel & contrivance of a voluntary Agent. . . ."

Newton was so committed to the purely mechanistic aspects
of the gravitational laws that he was obliged to invoke the intervention of God to cover the contingency that the planetary
system would need divine restoration should the irregularities
introduced by the action of comets and planets on one another
destroy the perfection of the motion.

> . . . Blind fate could never make all of the Planets move one
> and the same way in Orbs concentrick save in considerable
> Irregularities excepted, which may have arisen from the
> mutual Action of Comets and Planets upon one and anoth-
> er, and which will be apt to increase, till this System want
> a Reformation. . . ."[4]

At the time of Newton's death in 1727, informed opinion had
become almost wholly convinced that man existed in a Sun-
centered, infinite universe populated by an infinite number of
particles whose behavior was governed by the Newtonian laws
of inertia and gravity.

This idea of the infinite universe soon raised many issues
that led to great observational and theoretical contention.
Some have been resolved, but others remain with us today. In
at least three of these major cosmological problems the work

---

4. Sir Isaac Newton, *Opticks: or, a Treatise of the Reflexions, Refractions, Inflexions and
Colours of Light.* (London: S. Smith and B. Walford, 1704), query 31. This passage
occurs in the 2nd, 3rd, and 4th English editions.

of William Herschel occupies a central position in the tortuous path toward a new understanding.

William Herschel

Herschel was born in 1738, eleven years after the death of Newton, and he lived to be the first President of the Royal Astronomical Society, founded in 1820. His death two years later marked the end of an epoch during which, primarily through his efforts, the modern science of cosmology was created. He built the first telescopes capable of penetrating far into space, he compiled great catalogues of stars and objects called nebulae, and he speculated and formulated theories about the structure of the universe.

An extraordinary feature of Herschel's career is that until he was thirty-five years old there is little indication that he had any interest in astronomy. In his youth he was an oboeist in the Hanoverian Foot Guards, but at the age of nineteen he decided to leave the army and pursue a musical career in England. This he did with increasing success in various activities. Eventually, in 1766, he was appointed organist in Halifax. Almost simultaneously he was nominated as organist at the Octagon Chapel[5] in Bath and took up this appointment in December of that year. In Bath Herschel continued to pursue a successful musical career. In addition to his duties at the Octagon he was a member of the Pump Room orchestra and had many pupils. After he had lived in Bath for seven years, his notes give some indication of an interest in astronomy. On April 19, 1773, he records that he bought a quadrant and a book on trigonometry. Previously the only mention of astronomy in his notebook is that he had observed Venus and an eclipse of the Moon in 1766. The year 1773 seems to have

5. The Octagon Chapel, 53 feet in diameter and to the domed lantern roof 53 feet high, was opened in 1767 and remained a fashionable chapel for over one hundred years, being recommended to the notice of invalids as a church without drafts, dry, and as a place where good preaching could be heard. Herschel was the first organist. The building is now in the possession of the Corporation of the City of Bath and is used for various activities.

been the turning point in Herschel's interest. He bought many books on astronomy, hired a reflecting telescope, and decided to build one of his own, purchasing equipment to make speculum mirrors in September of that year. After experimenting with reflectors of the Gregorian type, he decided to make the Newtonian form of telescope, which he found easier to use.

Herschel created a foundry in his house in order to cast the blanks of speculum metal. In November 1778, using a composition of 29.4 percent tin and 70.6 percent copper, he cast and polished a very good mirror with a focal length of seven feet. This telescope soon made Herschel famous. When searching the sky during the night of March 13, 1781, he noticed a curious "nebulous star or perhaps a comet." After a few hours' observation he found that the position of the object changed with respect to the stellar background. When the news of this discovery reached Maskelyne, the Astronomer Royal, he was able to obtain confirmation of the existence of this object, and on April 26 Herschel presented a paper to the Royal Society, "Account of a Comet." Soon it was clear that the object discovered by Herschel was not a comet but a new planet. Herschel proposed to call the new planet *Georgium Sidus* in honor of the king, but later the name was changed to Uranus.[6]

Thus, at the age of forty-three, Herschel, still a professional musician, became a famous astronomer. He was elected a fellow of the Royal Society on December 6, 1781. Sir Joseph Banks, the President of the Royal Society, was so impressed with this work that he suggested to King George III that Herschel be relieved of the necessity to earn his living as a musician. The king summoned Herschel to Windsor and appointed him private astronomer to the king at a salary of £200 a year.

The Structure of the Milky Way

Anyone who contemplates the beauty of the heavens on a

6. Sixteen years later, in 1787, Herschel discovered two satellites of this planet.

clear dark night will see that a band of diffuse light—the
Milky Way—stretches around the sky. The realization that
this band of light could be resolved into many stars by the use
of a telescope conflicted with the idea of an infinite system of
stars distributed symmetrically around the Sun. This problem
of the "construction of the heavens" became a major preoc-
cupation of Herschel after his discovery of the planet Uranus
in March 1781. Through the patronage of George III,
Herschel was able to abandon his career as a musician in
Bath, and by 1782 he was installed at Datchet. He moved to
Slough four years later. In the twenty years following his arrival at Datchet, Herschel, with the devoted help of his sister
Caroline, made three remarkable surveys of the heavens. Although he constantly strove to build larger telescopes in order
to increase his power of penetrating into space, the major part
of these surveys was carried out with a telescope of 20 feet
focal length (aperture 18.7 inches). He first used this telescope
in October 1783, and his three major publications on the construction of the heavens, which appeared in 1784, 1785, and
1789,[7] are based largely on surveys made with this instrument.
His ambition to construct an even larger telescope eventually
resulted in the famous instrument of 40 feet focal length with
an aperture of 48 inches. He began observing with this instrument in 1787, but for various reasons was never able to obtain
results as significant as those he achieved with the 20-foot
telescope.[8]

Although Herschel's fame at this time rested on his discovery of Uranus, he had an earlier interest in the problem of
nebulae and the construction of the heavens derived from his

7. William Herschel, "An account of some Observations tending to investigate the
Construction of the Heavens,' *Phil. Trans. R. Soc. Land.* 74 (1784): 437–51. Read June
17, 1784. Idem, "On the Construction of the Heavens," *Phil. Trans. R. Soc. Lond.* 75
(1785): 213–66. Read February 3, 1785. Idem, "Catalogue of a second Thousand of
the new Nebulae and Clusters of Stars; with a few introductory Remarks on the
Construction of the Heavens," *Phil. Trans. R. Soc. Lond.* 79 (1789): 212–55. Read June
11, 1789.
8. A detailed account of Herschel's work on the construction of telescopes has been
given by J. A. Bennett, " 'On the power of penetrating into space': The telescopes of
William Herschel," *Jr. History Astronomy* 7 (1976): 75.

reading of Robert Smith's book *A Compleat System of Opticks,* published in 1738.[9] This work not only stimulated Herschel to begin the construction of his telescopes but also revealed to him the problems of stars and nebulae. In his first journal of 1774 he refers to observations of the nebula in Orion. It seems that in December 1781, nine months after the discovery of Uranus, his faithful friend in Bath, Dr. William Watson, presented him with a copy of Messier's catalog of 103 nebulae and star clusters.

Herschel's association with Watson had arisen accidentally two years earlier. In December 1779 Herschel was observing the Moon through a telescope he had placed in the street in front of his house when a stranger, William Watson, asked permission to look through the telescope. Soon they became close friends, and it was Watson—who had been elected a fellow of the Royal Society in 1767—who introduced him to the Bath Philosophical Society and communicated his early papers to the Royal Society. When Watson gave him the copy of Messier's catalog, Herschel realized that his own telescopes were superior to those available to Messier, and it was this which launched Herschel in October 1783 on his surveys of the heavens. From that time onward, Herschel's observations and publications are concerned inextricably with the problem of the structure of the Milky Way and the nature of nebulae. For simplicity of description we shall first consider the problem of the structure of the Milky Way, which historically appears as the major content of the second (1785) of Herschel's three papers on the construction of the heavens.

Before Herschel turned his attention to this problem, Thomas Wright had already proposed that stars were not distributed symmetrically around the Sun. In his book *An Original Theory, or new Hypothesis of the Universe*[10], published in 1750,

9. Robert Smith, *A Compleat System of Opticks* (Cambridge, 1738). Robert Smith (1689-
-1768) had been a professor of astronomy and experimental philosophy and master of Trinity College at the University of Cambridge.
10. London: 1750. H. Chapelle, A facsmile reprint together with the first publication of "A Theory of the Universe" (1734) with an introduction and transcription by M. A. Hoskin was published by MacDonald (London) and Elsevier (New York) in 1971.

Wright suggested a possible means of reconciling the infinite extent of stars with the appearance of the Milky Way as a band of light stretching around the sky. He proposed that the Sun was at the center of a giant disk extending infinitely outward. An observer on Earth would thus see only relatively few stars when looking perpendicularly to the disk, but would see their infinite extent forming the band of the Milky Way when viewing parallel to the disk. This hypothesis seems to be the first scientific argument that stars in the heavens are not arranged with spherical symmetry around the Sun. It was a concept preserving the priority of the Sun and a new kind of symmetry to be associated with a divine creator. A summary of Wright's hypothesis, published in a German periodical, formed the basis of Immanuel Kant's *Universal Nature History and Theory of the Heavens,* published in 1755.[11] These works were available thirty years before Herschel began his surveys of the heavens, but he makes no reference to them in his papers. In view of the similarity of his own conclusions, the question has been raised whether he knew of the existence of these hypotheses. In a recent study of Herschel, Hoskin states that Herschel certainly owned a copy of Wright's book, but Hoskin concludes that Herschel may have glanced through Wright's book when his own studies of the Milky Way were well advanced, "expecting to learn little from it, and that his expectations were fulfilled."[12] In any event, Herschel's conclusions were deduced from a vast amount of observational data not available to his predecessors.

The fundamental difficulty of reaching any scientific conclusions from a study of the stars in the eighteenth century was that no means existed for measuring their distances. When the disputes of the Copernican-Galilean era had subsided, astronomers soon appreciated that the motion of the Earth around the Sun provided them with a moving platform

11. Immanuel Kant, *Universal Nature History and Theory of the Heavens (Allgemeine Naturgeschichte und Theorie des Himmels)* (Königsberg and Leipzig: J. F. Petersen, 1755).
12. M. A. Hoskin, *William Herschel and the Construction of the Heavens* (London, 1963), p. 115.

over a huge baseline from which it should be possible to detect
the parallactic displacement of the nearer stars, and thus, by
simple trigonometry, to determine their distances. The op-
timism of early attempts to make this measurement rested on
a complete lack of appreciation of the actual distance of the
stars, and long before Herschel faced the problem many observers had been defeated. In order to place the issue in perspective it may be mentioned that the parallactic displacement of even the nearest star to the Sun corresponds to an
image displacement of only 0.03 mm at the focus of a telescope
with a focal length of 4 meters when observing the star across
the diameter of the Earth's orbit. Such precision lay far
beyond the resources available to Herschel or any other
astronomer in the eighteenth century, although attempts to
measure the parallax led accidentally to other major discoveries.

Herschel certainly believed that with his superior telescopes
he could derive a method for measuring the parallax. He
thought he could overcome many of the difficulties by measurements on double stars—pairs of stars—which at that time
were believed fortuitously to appear close together. His plan
was to make precision measurements of the separation of
those close pairs throughout the year in the expectation that
the parallactic shift of one of the stars would be greater than
that of its fainter companion, believed to be more distant. In
1782 he explained his method to the Royal Society[13] and
tabulated his first catalogue of 269 double stars. Immediately
disputes arose because of his assumption that the stars were
about the size and luminosity of the Sun and that the difference in their apparent magnitudes was entirely the result of
their differing distances. In their comments on the paper,
Maskelyne and the Royal Society Committee wrote that in
their view the postulate was positively false. Herschel was referred to a paper published some years earlier by the Rever-

13. William Herschel, "On the Parallax of the Fixed Stars," *Phil. Trans. R. Soc. Lond.*
72, (1782): 82–111. Communicated by Sir Joseph Banks, Bart., P.R.S. Read December 6, 1781.

end John Michell[14] in which attempts to estimate the varying brightnesses of stars led Michell to disagree with the standpoint adopted by Herschel that the apparent magnitude could be linearly related to the distance of a star.

Michell's paper contained an even more fundamental objection to Herschel's view that double stars were fortuitously chance groupings of a near and a distant star. Nearly fifty years earlier, Edmond Halley had tested the prediction that stars are distributed at regular intervals throughout space, and had been able to make estimates of the increases in the number of stars as the distance from the Sun increased. On the basis of Halley's data, Michell had computed the number of *chance* groupings of two stars, of the type being investigated by Herschel, and had concluded that the odds against double stars being accidental groupings were overwhelming.

After the publication of Herschel's papers, Michell communicated a note to the Royal Society in which he wrote: "It is not improbable that a few years may inform us, that some of the great number of double, triple, stars etc. which have been observed by Mr. Herschel, are systems of bodies revolving around each other."[15] Michell's view was correct. More than twenty years later, when Herschel's major surveys of the heavens had been completed, he returned to the measurement of double stars and found, as Michell had predicted, that their relative positions had changed. Herschel's attempt to use double stars for parallactic measurements had been doomed to failure, but he had by accident provided the first observational evidence for the existence of binary stars moving around one another under the inverse square law of Newtonian gravitation.[16]

14. John Michell, "An inquiry into the probable Parallax, and Magnitude of the fixed Stars, from the Quantity of Light which they afford us, and the particular Circumstances of their Situation," *Phil. Trans. R. Soc. Lond.* 57, (1767): 234–64. Read May 7 and 14, 1767. John Michell lived from 1724 to 1793. He was the inventor of the torsion balance and was elected a fellow of the Royal Society in 1760.
15. *Phil. Trans. R. Soc. Lond.* 74 (1784): 56.
16. Herschel's accounts of these changes in relative position and his conclusion that the double stars were binary systems were published in three papers: "Catalogue of 500 new Nebulae, nebulous Stars, planetary Nebulae, and Clusters of Stars; with

Herschel's work on double stars—eventually he cataloged 848–although failing in the prime intention of measuring distances, had another important outcome. He had suggested in his original parallax paper of 1782 that measurements on these stars could be used to detect any motion of the observer through space in which the double stars did not share—in other words, the motion of the solar system through space. Fundamentally the ideas were not new; the difficulty was to disentangle the true proper motions of stars through space from the apparent motion caused by the movement of the solar system. Over many years, extending almost to the time of his last published papers, Herschel pursued this problem and answered many criticisms. His conclusion that the Sun is moving toward the star λ Herculis agrees within the limits of uncertainty of modern measurements.

The discovery of true binary stars and the evaluation of the direction of the solar motion were effectively by-products of the untenable idea that double stars could be used for parallactic measurements. In his work on the structure of the Milky Way Herschel persisted in his assumption that the brightness of a star could be equated with nearness. It is true that the modern astronomer frequently uses this principle for measuring the distances of stars from a comparison of their apparent brightness, but stars used for this purpose are from categories carefully selected on the basis of their spectral type or other characteristics, which gives confidence that all stars in such a category are of the same absolute brightness. Today we know, as Herschel did not, that over the totality of stars there is an enormous variation in inherent brightness—some are a million times brighter than the Sun and others a thousand times less bright. Bearing this in mind, it seems the more re-

Remarks on the Construction of the Heavens," *Phil. Trans. R. Soc. Lond.* 92(1802): 477–528. Read July 1, 1802. "Account of the Changes that have happened, during the last Twenty-five Years, in the relative Situation of Double-stars; with an investigation of the Cause to which they are owing," *Phil. Trans. R. Soc. Lond.* 93(1803): 339–82. Read June 9, 1803. "Continuation of an Account of the Changes that have happened in the relative Situation of double stars," *Phil. Trans. R. Soc. Lond.* 94(1804): 353–84. Read June 7, 1804.

markable that Herschel was able to make any progress with conclusions based on an assumption that was wrong by a factor of a billion.

Herschel called his method *Gaging the Heavens,* or the *Star-Gage.* It was a pioneer work in the collection and interpretation of statistics on the stars—a mammoth task pursued relentlessly on every clear night and carried out under primitive conditions with his sister Caroline serving as his scribe. His technique was to count the number of stars in ten fields of view of his telescope, which were very near each other, and thereby obtain a mean count of the number in that particular direction of the heavens. For the distribution about the Sun he constructed ray paths of lengths equivalent to the gages in that direction; the terminating points of these paths then delineated the boundary of the sidereal stratum.

In his second paper on the construction of the heavens (1785) Herschel wrote of the Milky Way that he had "now viewed and gaged this shining zone in almost every direction, and find it composed of stars whose number, by account of these gages, constantly increases in proportion to its apparent brightness to the naked eye." In this historic paper Herschel entered into a detailed analysis of the gages and their translation to the delineation of the structure of the Milky Way, and remarked, "In the most crowded part of the Milky Way I have had fields of view that contained no less than 588 stars and these were continued for many minutes, so that in one quarter of an hour's time there passed no less than 116,000 stars through the field of view of my telescope." He described the Milky Way as a nebula of "a very extensive, branching, compound Congeries of many millions of stars." He remarked that all the stars that could be seen with the unaided eye would be contained within a very small sphere drawn around the central star, which he assumed to be the Sun. Herschel's scheme for the structure of the Milky Way bore some similarity to that of Thomas Wright with the important distinction that, whereas in Wright's scheme the disk extended infinitely, Herschel concluded that the stars were contained in an enclosure somewhat similar in shape to a flat rectangular box having a length five times its thickness.

Herschel's achievement in deriving even an approximate structure for the Milky Way from the principle of star gaging is thrown into relief when we consider that no further significant progress was made for nearly 150 years. The conclusions derived from his star gaging destroyed the idea of spherical symmetry for the Milky Way and established an approximate shape for the system. It is strange that although Herschel cataloged the globular clusters and must have been well aware of their asymmetrical distribution around the Sun, neither he nor any astronomer of the next century recognized the inescapable significance of this asymmetry until Shapley did so 140 years after the publication of Herschel's measurements.

The Problem of the Nebulae

The problem of the nebulae, or the regions of apparently luminous clouds to be seen in the Milky Way, had been a matter of contention long before Herschel turned his attention to it. Many regions that appear as luminous clouds to the unaided eye are resolved into stars by small telescopes. The demonstration of this by Galileo early in the seventeenth century supported the belief that all such nebulae could be resolved, given a sufficiently powerful telescope. Contrasting opinions before Herschel began his studies were represented by Edmond Halley, who believed that the nebulae shone with their "own proper lustre," and by Michell, who maintained that the nebulae were composed of stars and that those nebulae "in which we can discover either none, or only a few stars, even with the assistance of the best telescopes, are probably systems, that are still more distant than the rest." Michell's opinion was based on a statistical investigation of the dimensions and brightness of the unresolved nebulae and rested on a firmer scientific basis than the more widely known views of

Thomas Wright, Immanuel Kant, and Lambert,[17] all of whom promulgated the view that the nebulae were star systems probably remote from the Milky Way.

By the mid-eighteenth century, catalogues of the nebulae were appearing in print. In 1755 La Caille[18] published a list containing 42 nebulae; Messier's catalogs of 103 objects, still identified by his name, appeared in 1780 and 1781.[19] The presentation of Messier's catalog to Herschel by William Watson stimulated him to his surveys of the heavens with the 20-foot reflector. On June 17, 1784 when he read the first of his three papers on the *Construction of the heavens* to the Royal Society, Herschel was clearly excited by the power of his telescope:

> The excellent collection of nebulae and clusters of stars which has recently been given in Connaissance des Temps for 1783 and 1784, leads me next to a subject which, indeed, must open a new view of the heavens. As soon as the first of these volumes came to my hands, I applied my former 20 feet reflector of 12 inches aperture to them; and saw, with the greatest pleasure, that most of the nebulae, which I had an opportunity of examining in a proper situation, yielded to the force of my light and power, and were resolved into stars.

The superiority of Herschel's telescopes over any previously used is clearly indicated by this later passage:

> When I began my present series of observations, I surmised, that several nebulae might yet remain undiscovered, for want of sufficient light to detect them; and was, therefore, in hopes of making a valuable addition to the clusters of stars and nebulae already collated and given us in the work

17. Johann H. Lambert, *Cosmologische Briefe über die Einrichtung des Weltbaues* Bey e Kletz wittib., (Augsberg: 1761).
18. Nicholas de la Caille, (also known as Lacaille), "Sur les étoiles du ciel austral," *Hist. Acad. R. Sci. Paris*, 1755.
19. Charles Messier, "Catalogue des Nebuleuses et des amas d'Étoiles," *Connais. Temps (Paris)* 1783 (1780): 225–49. This catalog contained 68 objects. "Catalogue des Nebuleuses et des amas d'Étoiles" *Connais. Temps (Paris)* 1784 (1781): 227–69. The full list of 103 objects is found in this reference.

before referred to, which amount to 103. The event has plainly proved that my expectations were well founded; for I have already found 466 new nebulae and clusters of stars, none of which, to my present knowledge, have been seen before by any person, most of them, indeed, are not within the reach of the best common telescopes now in use.

Eventually Herschel's catalog of nebulae ran to 2500 objects. His confidence that all were resolvable into stars quickly led him to maintain that nebulae were collections of stars in the manner of the Milky Way, but existing as separate external systems. Although it is debatable whether Herschel's confidence was completely maintained in this matter, there is no question that he was soon forced to divide nebulae into various forms. There were two particular difficulties. In the second survey paper he referred to a number of objects, such as the nebula in Orion, which he found impossible to resolve into stars, but concluded that they must be very distant stellar systems which "cannot be otherwise than of a wonderful magnitude, and may well outvie our Milky Way in grandeur." An even greater difficulty presented itself when he discovered "planetary nebulae": "a few heavenly bodies, that from their singular appearance leave me almost in doubt where to class them." On February 10, 1791, he reported to the Royal Society:

November 13, 1790. A most singular phenomenon! A star of about the 8th magnitude, with a faint luminous atmosphere, of a circular form, and of about 3 min. of arc in diameter. The star is perfectly in the center, and the atmosphere is so diluted, faint and equal throughout, that there can be no surmise of its consisting of stars; nor can there be any doubt of the evident connection between the atmosphere and the star. [He was no longer able to maintain his confidence in the resolvability of all nebulae, for he had found a star] involved in a shining fluid, and a nature totally unknown to us. . . . More extensive views may be derived from this proof of the existence of a shining matter.

Perhaps it has been too hastily surmised that all milky nebulosity, of which there is so much in the heavens, is owing to starlight only. These nebulous stars may serve as a clue to unravel other mysterious phenomena.[20]

In the light of our present knowledge we cannot but have great admiration coupled with feelings of sympathy for the vast task that faced Herschel. He had penetrated far enough into space and with sufficient resolving power to reveal phenomena that could have no explanation within the scope of the astronomy and physics of the eighteenth and nineteenth centuries. Today we know that the use of the term nebulae to cover the 103 objects in Messier's catalog (which stimulated Herschel's surveys) includes a range of completely different astronomical bodies. There are, as Herschel believed, objects that are resolvable into stars, the extragalactic nebulae. But the catalog also lists globular star clusters (now known to be within the framework of the Milky Way), true gaseous nebulae such as the Orion nebula, and the gaseous nebulae of supernova remnants in the Milky Way. These various features were not finally clarified until 1926.

An outstanding practical effort to pursue Herschel's concepts was made by the third Earl of Rosse at Birr in Ireland. Herschel's 48-inch telescope had never performed satisfactorily, but in 1845 Rosse brought into use a reflector with an aperture of 72 inches, which he had built. Like Herschel, he failed to resolve the Orion nebula into stars, but he found that many other nebulae were resolvable and exhibited a marked spiral structure. No means of estimating distances were available, so Rosse did not reach the correct solution that they were outside the Milky Way but considered these spirals to be clusters of stars within the Milky Way.

The opinion that any nebulae were extragalactic star systems received a severe setback in 1863 when William Huggins used the new spectroscopic techniques to investigate nebulae.

20. William Herschel, "On Nebulous Stars, properly so called," *Phil. Trans. R. Soc. Lond.* 81 (1791): 71–88. Read February 10, 1791.

In an evening discourse at the Royal Institution on May 19, 1865, he said:

> Besides the stars, the heavens are mottled over with feebly shining cloud-like patches and spots, often presenting strange and fantastic forms. Between 5000 and 6000 of these so-called *Nebulae* are known. What is the nature of these strange objects? Dense swarms of suns melted into one mass by their enormous distance? Chaotic masses of the primordial material of the Universe? The telescopes alone would fail to give the answers to these questions and the analysis by the prism of objects so feebly luminous appeared hopeless. In August last, the speaker directed his telescope armed with the spectrum apparatus, to a small but comparatively bright nebula. His surprise was great to observe, that in place of a band of coloured light, such as the spectrum of a star would appear, the light of this object remained concentrated in three bright, bluish-green lines, separated by dark intervals. This order of spectrum showed the source of the light was luminous *gas*.[21]

It is ironic that the conclusions of Rosse, who had observed the nebulae with the largest telescope in the world, and the work of Huggins, who had fortuitously chosen a nebula that actually was luminous gas within the confines of the Milky Way, seemed to provide conclusive evidence against the extragalactic concept and thus seemed to destroy Herschel's major conclusions. At the end of the century Agnes Clerke, a distinguished historian, wrote that the question "hardly any longer needs discussion—no competent thinker with the whole of the evidence before him, can now, it is safe to say, maintain any single nebula to be a star system of co-ordinate rank with the Milky Way."[22]

21. Reprinted in *Royal Institution Library of Science: Astronomy* (London: Elsevier, 1970), 1:42.
22. Agnes Clerke, *The System of the Stars* (London: Black, 1890). This book and all subsequent editions for the next fifteen years maintained the view that no single nebula was a star system of coordinate rank with the Milky Way.

When these issues were finally settled a century after
Herschel's death, it was realized that everyone had been correct in some respects during the previous centuries; some
nebulae could be resolved into stars and were extragalactic,
others could never be resolved and existed as truly gaseous
nebulosities within the Milky Way system. Of all the protagonists in this long struggle, Herschel, in his final papers,
came closest to the truth in his recognition that not one but
several classes of nebulae existed.

The Evolutionary Problem

During Herschel's life the powerful legacies of Newton's
outlook on the mechanism of the universe prevailed. Nowhere
is the greatness of Herschel more evident than in his introduction of the ideas of evolution into the universe. He found it
necessary to divide the nebulae into several classes, and toward the end of his second paper on the *Construction of the heavens* he wrote:

> If it were not perhaps too hazardous to pursue a former
> surmise of a renewal in what I figuratively call the Labora-
> tories of the Universe, the stars forming these extraordinary
> nebulae, by some decay or waste of nature, being no longer
> fit for their former purposes, and having their projectile
> forces, if any such they had, retarded in each other's at-
> mosphere, may rush at last together, and either in suc-
> cession, or by one general tremendous shock, unite into a
> new body. Perhaps the extraordinary and sudden blaze of a
> new star in Cassiopeia's chair, in 1572, might possibly be of
> such a nature.

Here we find an insight suggesting that physical processes
other than the forces of gravitational attraction might be con-
ditioning the nature of the universe. The implication that
changes in the universe could occur because of stellar col-

lisions or deceleration in the atmospheres of celestial objects was a foresight of processes whose significance has only been realized in our own age. Herschel continued to elaborate his theory that nebulae were collections of stars developing and clustering under the action of attractive powers:

> This method of viewing the heavens seems to throw them into a new kind of light. They now are seen to resemble a luxuriant garden, which contains the greatest variety of productions, in different flourishing beds; and one advantage we may at least reap from it is, that we can, as it were extend the range of our experience to an immense duration. For, to continue the simile I have borrowed from the vegetable kingdom, is it not almost the same thing, whether we live successively to witness the germination, blooming, foliage, fecundity, fading, withering and corruption of a plant, or whether a vast number of specimens, selected from every stage through which the plant passes in the course of its existence, be brought at once to our view?

In his discussion of the objects he called planetary nebulae (i.e., stars enveloped in unresolved nebulosities) he correctly foresaw the modern theories of star formation in gaseous nebulae: "If, therefore, this matter is self luminous, it seems more fit to produce a star by its condensation than to depend on the star for its existence."

The temporal concept Herschel introduced as vital to our understanding of the universe must be counted as one of his greatest achievements. In a paper written when he was seventy-six years old the following passage occurs in referring to the Milky Way:

> ... for the state into which the incessant action of the clustering power has brought it at present, is a kind of chronometer that may be used to measure the time of its past and future existence; and although we do not know the rate of going of this mysterious chronometer, it is never-

theless certain, that since the breaking up of the parts of the Milky Way affords a proof that it cannot last for ever, it equally bears witness that its past duration cannot be admitted to be infinite.[23]

Perhaps his finest passage was written twelve years earlier. He correctly foresaw and expressed with precision the temporal concept that is so fundamental to the cosmology of the twentieth century:

> . . . a telescope with a power of penetrating into space, like my 40-feet one, has also, as it may be called, a power of penetrating into time past. To explain this, we must consider that, from the known velocity of light, it may be proved that when we look at Sirius, the rays which enter the eye cannot have been less than 6 years and 4½ months[24] coming from the star to the observer. Hence it follows, that when we see an object of the calculated distance at which one of these very remote nebulae may still be perceived, the rays of light which convey its image to the eye, must have been more than nineteen hundred and ten thousand, that is, almost two millions of years on their way; and that, consequently, so many years ago, this object must already have had an existence in the sidereal heavens, in order to send out those rays by which we now perceive it. . . .

Apart from this series of observations of the heavens, Herschel discovered the existence of infrared radiation from the Sun in 1800 and thereby opened to observation the first extension of the electromagnetic spectrum beyond the visible region.[25] He died in 1822 after a career as musician and astronomer in which he had arrived at a variety of conclusions

23. William Herschel, "Astronomical Observations relating to the sidereal part of the Heavens, and its Connection with the nebulous part; arranged for the purpose of a critical Examination," *Phil. Trans. R. Soc. Lond.* 104 (1814): 248–84. Read February 24, 1814.
24. The modern figure is 8.7 years.
25. An account of this discovery and of Herschel's life may be found in a paper by R. V. Jones, "Through Music to the Stars—William Herschel 1738–1822, *Notes and Records of the Royal Society* 33 (1978): 37.

about the universe, most of which were only generally recognized to be correct a century after his death.

10

The Size of the Universe:
The Earth and
the Planetary System

Determination of the size of the Earth and its distance from
the planets and stars have been issues of profound difficulty in
the evolution of cosmological ideas. In many instances even-
tual success in these measurements, particularly of the distance of stars, first in the Milky Way and then in nebulae, has
led to revolutions in cosmological thought. Even the relatively
local problem of distance measurements in the solar system
has been surmounted only in recent years. The fundamental
starting point for all these measurements concerns the size of
the Earth.

The Size and Shape of the Earth

The size of the Earth was discussed by Aristotle in his
Meteorologica. One of his pupils, Dicaearchos of Messina, who
lived from 350 to 290 B.C., was a prominent Greek geographer,
but his figures for the size of the Earth are gross under-
estimates. There are no records of actual attempts to measure
the Earth's circumference until that made by Eratosthenes of
Cyrene in the third century B.C. He was summoned from
Athens to Alexandria by Ptolemaios III Euergetes to educate

the future Ptolemaios IV Philopator and take charge of the great library collection. Eratosthenes excelled in many disciplines, but few of his scientific writings have survived. We are indebted to Cleomedes[1] for a description of the method by which Eratosthenes measured the circumference of the Earth —a method whose principle has survived to modern times.

Eratosthenes estimated the arc of the great circle through Alexandria and Syene (Aswan). The distance between these cities had been measured by Egyptian surveyors, and it was known that at noon on the day of the summer solstice the Sun was directly overhead at Syene because it did not cast a shadow. Eratosthenes therefore erected a vertical gnomon at Alexandria and measured the angle the rays of the Sun made with the vertical when it was in the zenith at Syene. A simple geometrical construction shows that this angle must equal the angle subtended at the center of the earth by the line joining the two cities. He measured the angle to be 7.2 degrees, or one fiftieth of a full circle. Hence the distance between Alexandria and Syene must be one fiftieth of the circumference of the Earth. The surveyors had found that Alexandria and Syene were separated by 5000 stadia; hence, the circumference of the Earth was calculated to be 250,000 stadia. As Eratosthenes probably used the Egyptian stadia of 516.73 feet, this figure is equal to 24,662 miles. This is within 2 percent of the modern value of 24,989 miles for the equatorial circumference of the Earth.[2]

Little improvement on this measurement of Eratosthenes was made for nearly 2000 years. In the last century B.C. Posidonius of Apamea considered that he had improved on this measurement; he used as a baseline the distance between Rhodes and Alexandria determined on the basis of a ship's sailing time as 5000 stadia. This is 25 percent too high, but

1. A popular writer on science in the Graeco-Roman world, of uncertain date, but probably early second century A.D.
2. This impressive accuracy is somewhat fortuitous because there are two compensating errors in the computations of Eratosthenes. The distance between Syene and Alexandria is 5346 Egyptian stadia, not 5000, but this error is compensated by the fact that Eratosthenes assumed that the two cities were on the same meridian.

Posidonius also made an error in measuring the difference between the angle of the Sun's rays at Alexandria and Rhodes.
This should have been 5.25 degrees, but Posidonius measured
the angle as 7.5 degrees, or one forty-eighth of a great circle.
The result, as reported by Cleomedes, was 240,000 stadia—
23,546 miles—about 6 percent less than the modern value for
the equatorial circumference, again a fortunate cancellation of
two errors of measurement. In the ninth century A.D. attempts
were made in Arabia to measure the size of the Earth. The
Caliph al-Ma'mūn (A.D. 786–833) founded the Baghdad
Academy of Science and encouraged many scientific en-
terprises, among them the attempt to measure the Earth's circumference by setting out a baseline measured by wooden
rods on the Zinjar plateau close to Baghdad. The angles of the
Sun's rays were also measured more accurately than before,
and the value for the circumference was only 3.6 percent too
great. This is probably the first accurate measurement, as distinct from the fortuitous cancellation of two errors that accidentally gave reasonable results for Eratosthenes and
Posidonius.

Later, in the Middle Ages, gross errors were introduced.
The most important error arose from a theological solution of
the problem that had remarkable consequences. Cardinal
Pierre d'Ailly[3] thought that the size of the earth could be derived from information contained in the Apocrypha. The sixth
chapter of the second book of Esdras summarizes the works of
Creation.[4] Cardinal d'Ailly considered these statements to be
of undisputed authority, and argued that since only one sev-

3. Pierre d'Ailly, born at Campiegne in Picardy in 1350, was a celebrated French
prelate, metaphysician, and divine. In 1384 he became *grandmaître* of the college of
Navarre, Paris, where his eloquent lectures attracted large crowds. He became chan-
cellor of the University of Paris in 1389 and was made a cardinal in 1411 by the
antipope John XXIII (Baldassare Cossa). He died about 1425.
4. The important passages are are v.42: "Upon the third day thou didst command
that the waters should be gathered in the seventh part of the Earth, six parts hast thou
dried up, and kept them, to the intent that of these some being planted of God and
tilled, might serve thee" and v. 47: "Upon the fifth day thou saidst unto the seventh
part, where the waters were gathered, that it should bring forth living creatures, fowls
and fishes: and so it came to pass."

enth of the Earth's surface was water, the ocean between Europe and the east coast of Asia could not be very wide.[5] Because d'Ailly thought he knew the extent of the land, he concluded that the earth must be much smaller than currently believed. He stressed this in his book *Imago mundi,* an edition of which came into the hands of Columbus when he was planning his westward voyages. Columbus obtained support on the evidence in this work that the passage across the ocean to Marco Polo's land of Zipango in Asia was short. It is a curious fact that had Columbus based his estimates on the nearly correct Greek values for the size of the Earth, he is unlikely to have obtained support, or indeed attempted the voyage that destroyed nearly all geographical conceptions based on sacred writings.[6]

At last, in the seventeenth century, contemporary methods of observation, based on the principle of Eratosthenes, gave a new order of precision to the measurement of the size of the Earth. Baselines were determined by triangulation using theodolites, and telescopes were used to observe the angle of a star from the zenith as it crossed the meridian. Christian Huyghens, born in 1629 in Holland, made many improvements of the lenses used in telescopes and their performance was greatly superior to that of the instruments used by Galileo. In his early thirties Huyghens invented a new eyepiece with two lenses for use with telescopes. He was also the first to construct a pendulum clock. Such a precise development soon led to the need for some means of making accurate measurements of the direction in which a telescope was pointing. Francesco Generini in France, William Gascoigne[7] in England, and Geminiano Montanari of Bologna almost simulta-

5. The general belief was that the Orient would be reached by sailing westward—an opinion shared by Columbus. When he sighted land on October 12, 1492 (the Bahamas), he was convinced that he had reached the East Indies.
6. A copy of d'Ailly's book *Imago mundi* annotated by Columbus is in the library of Seville. A letter from Columbus acknowledging his indebtedness to the mistake based on the book of Esdras is in *Viajes y descubrimientos* by Navarrete (Madrid 1825) 1: p. 242, 264.
7. William Gascoigne was killed in the battle of Marston Moor, July 2, 1644.

neously (about 1640) fitted a thread micrometer to a telescope in order to measure the angular distances between neighboring stars. Then, in France, Jean Picard of the Paris Observatory added a graduated circle to the telescope.

Picard was born in 1620 and with Adrien Auzout, also of Paris, was a poineer in the attachment of the telescope to astronomical measuring instruments. Picard was the first to use the combination of an astronomical measuring instrument attached to a telescope, together with an accurate timepiece to observe the time of transit of a star across the meridian. He thereby introduced the modern method of determining the Right Ascension of stars. In northern France in 1671, Picard used this technique to make the first precision determination of an arc of the meridian, and hence of the size of the Earth, Picard's measurement enabled Newton to remove an inconsistency that appeared when he used the previously existing measurements to apply the inverse square law to the motion of the Moon and the fall of a body at the Earth's surface.

Seven years earlier, Picard had called the attention of Louis XIV to the lack of astronomical instruments in his kingdom, thus stimulating the king to erect the buildings that soon became famous as the Paris Observatory. In 1668 Louis XIV invited Cassini to Paris. Giovanni Domenico Cassini was the first of the Cassinis who influenced the work of the Observatory over nearly two centuries. He was born in 1625 in Perinaldo, Italy. At twenty-five he was appointed professor of astronomy at the University of Bologna. He remained there for nineteen years and distinguished himself in many activities apart from his work as an astronomer, especially in hydraulics and military problems. Summoned by Louis XIV to Paris, he arrived there in April 1669 when the Observatory was under construction. Until he became blind in 1710 (he died in 1712) he dominated the work of the Observatory. He was followed by his son, Giacomo (Jacques) Cassini (1677–1756). One of Jacques's three sons, César François (born in 1714), was appointed the first director of the Observatory in 1771.[8] The extension of the measurements of Picard to an arc of a meridian through the entire kingdom was among the many important observations carried out by the first Cassini. With the help of many collaborators, he established the "meridian of France" and published these new and accurate determinations at the age of seventy-five in *On the Magnitude and Form of the Earth*. This work was continued by Giacomo, who in 1718, assisted by Maraldi and de la Hire, measured the arc between Dunkirk and Montdidier and thus completed his father's work on the meridian of France.

During this period, difficulties arose because of inconsistencies in the various values obtained for the size of the Earth. In 1673 Huyghens in his *Horologium oscillatorium* defined the relationship between the time of oscillation and length of a simple pendulum: the time of oscillation is equal to the square root of the length divided by the acceleration due to gravity. Thus, if the Earth were a perfect spheroid, the gravitational attraction would everywhere be uniform, and the length of a pendulum, which oscillates with a period of a second, would everywhere be the same. In 1671 the French Academy organized an expedition to Cayenne in French Guiana. At this latitude of 5 degrees it was found that the length of a pendulum had to be shortened compared with that of a pendulum in Paris (at latitude 49 degrees) in order for a clock to keep the same time. These results, published in 1684, together with the measurements of the meridian arc by Picard, were used by Newton in his investigation of the shape of the Earth. They led to his conclusion that the Earth must be an oblate spheroid, that is, bulging at the equator and flattened at the poles. The explanation of the Cayenne pendulum measurements was then straightforward, namely, that the gravitational attraction was

8. Until 1771 all astronomers at the Observatory were Academicians and considered of equal rank. A fourth generation Cassini (Jean Dominique, 1748–1845) was sent to sea at age twenty to test marine chronometers. Eventually he returned to the Observatory and was engaged largely on administrative duties. He was arrested in 1794 as an aristocrat but was eventually released from prison and lived until the age of ninety-seven.

less near the equator because the pendulum was farther from the center of the Earth and therefore had to be shortened to beat the same time as the pendulum in Paris.

This result differed from that obtained by the Cassinis in France from the measurement of the meridian arc and from various measurements using a pendulum. They concluded that the Earth must be a prolate spheroid, that is, flattened at the equator and bulging at the poles. The issue of the shape of the Earth became a controversial matter, and in 1735 the French Academy of Sciences decided to settle the question. They dispatched expeditions to measure two meridian arcs, one at the equator and the other at the highest accessible latitude. After ten years of work, an arc of 3 degrees was measured close to the equator in Peru, and an arc of 1 degree was measured close to the Arctic Circle in Lapland.[9] The results left no doubt that the length of the meridian arc increased with increasing latitude, thus confirming that the Earth was flattened at the poles and that the shape was that of an oblate spheroid—the solid figure derived by rotating an ellipse about its minor axis. This result, in agreement with the calculations of Newton, was soon accepted.

For over a century after this controversy had been settled, knowledge of the Earth's size and shape was refined by successive improvements in measurement techniques. Theories had been developed to account for the Earth's shape. An Earth at rest would probably acquire a perfectly spherical shape, but the Earth rotates over a period of twenty-four hours, and under the combined influence of the gravitational and centrifugal forces, the sphere is flattened into an oblate spheroid. Precise calculations had been made on the assumption that the Earth has a plastic interior and that it was in

9. The Peruvian measurements were made by Pierre Bouguer, Charles Marie de la Condamine, and Louis Goden. The expedition to Lapland was led by Pierre Louis Moreau de Maupertius. The results were published in the classic work of Bouguer, *La Figure de la terre*. A member of the de Maupertius expedition was Alexis Claude Clairaut, a distinguished member of the French Academy who, in addition to his work on the shape of the Earth, calculated the perturbations of Halley's comet by the major planets and predicted the date of the return of the comet to within a month.

hydrostatic equilibrium, with the result that it was predicted to have a flattening ratio of 1/299.8 (i.e. the fractional difference between the equatorial and polar radii). The most accurate determinations derived from surface gravity measurements, geodetic studies, and the motion of the Moon gave a value of 1/297.1. This implied that the Earth was flattened at the poles so that the polar diameter was 7899 miles, but the equatorial diameter was 27 miles greater.

The launching of artificial Earth satellites in 1957 immediately provided a better method for measuring the flattening. The equatorial bulge of the Earth exerts a torque that causes the satellite orbit plane to precess about the polar axis. Accurate measurements of the rate of precession for the second Soviet satellite, Sputnik II, launched in November 1957, indicated that the flattening was about 1/298.2, implying that the difference between the polar and equatorial diameters was some 500 feet less than had previously been thought. The American *Vanguard I* satellite, launched in March 1958, enabled more accurate measurements of the flattening to be made, and geophysicists were forced to conclude that the Earth was not in hydrostatic equilibrium but that the interior had sufficient mechanical strength to support those stresses at the base of the mantle that would be associated with these departures from the equilibrium condition. More surprises soon followed because detailed analyses of satellite observations showed that there were other departures from the oblate spheroid: the Earth was "pear shaped," having a depression of 30 meters at the South Pole and a hump of 10 meters at the North Pole. These irregularities indicate that large shearing forces exist in the Earth's interior. For example, the hump at the North Pole represents forces sufficient to draw up the sea level by about 10 meters over an area comparable with the size of the Atlantic Ocean. [10]

10. Other depressions and humps in the shape of the Earth have been revealed. For example, there is a depression of 113 meters south of India and a hump of 60 meters near Great Britain. For a general account of this work, see D. G. King-Hele, *Satellites and Scientific Research* (London: Routlege & Kegan Paul, 1960), p. 91; and *Qu. Jr. Roy. Astr. Soc.* 13 (1972): 374.

Three hundred years after the arguments about whether the Earth was a prolate or an oblate spheroid, the Earth's shape is now known to within 10 meters. We are, indeed, far from the age when St. Augustine, although stressing the uses of science, was forced to deny the existence of the antipodes, or when Basil of Caesarea[11] declared that it was a matter of no interest "whether the Earth is a sphere or a cylinder or a disk, or concave in the middle like a fan."

Distances to the Sun, the Moon and Planets

During the second century B.C., Aristarchus of Samos was probably the first astronomer to attempt the actual measurement of the distances to the Moon and the Sun. For several centuries there had been a number of speculations about the magnitude of these distances. For example, Eudoxus, on the basis of Pythagorean musical intervals, had concluded that the Sun's diameter was nine times that of the Moon. The work of Aristarchus *On the Sizes and Distances of the Sun and the Moon* is a fine example of the geometrical methods that formed the basis of Greek astronomy. When the Moon is precisely half full, the angle subtended by the Earth and the Sun at the Moon is exactly a right angle. At this phase Aristarchus measured the angle between the Moon and the Sun. This angle determines the shape of the right-angled triangle whose vertices are the Moon, the Earth, and the Sun, and hence the ratio of distances to the Moon and the Sun.

Aristarchus measured the appropriate angle as 87 degrees, which gives the ratio of the distance to the Sun and the Moon as 19 to 1. He also knew from observations during an eclipse that the Sun and the Moon subtend the same angle at the Earth. Hence Aristarchus concluded that the Sun must also be 19 times as large as the Moon. Although the principle used

11. St. Basil the Great, born at Caesarea in Cappodocia A.D. 329, was the originator of monastic groups and discipline in the Eastern Church and a person of immense practical piety. The date of his death is believed to be 379.

by Aristarchus is perfectly sound, it is difficult to measure the relevant angle at the half-moon; not only is it difficult to determine the center of the Moon and the Sun but also he had no means of knowing when the Moon was exactly half full. Modern measurements give the relevant angle as 89 degrees 51 minutes, rather than the 87 degrees he measured. This makes a large difference in the ratio of the distances to the Sun and the Moon, which is 400 to 1 and not 19 to 1 as determined by Aristarchus.

In his conversion of these ratios to actual distances and sizes, Aristarchus was even more ingenious. The maximum duration of a lunar eclipse occurs when the Moon lies on the ecliptic and therefore passes through the center of the Earth's shadow. He compared the time for which the Moon was totally obscured with the interval between the instant when the Moon first touched the Earth's shadow and the instant when it became totally obscured. He discovered that these two intervals were equal. He thereby concluded that the breadth of the Earth's shadow at the distance of the Moon was twice the diameter of the Moon. Aristarchus accepted the measurements of Eratosthenes for the size of the Earth, and then, from a geometrical construction, calculated the actual distances (in stades) to the Moon and the Sun and their diameters. He concluded that the diameter of the Moon was 0.36 that of the Earth (the modern figure is 0.27), and the diameter of the Sun was 6.75 times that of the Earth (it should be 108.9). He measured the distance of the Moon as 9.5 Earth diameters (it is 30.2) and the distance of the Sun as 180 Earth diameters (it is 11,726). Although Aristarchus seriously underestimated the sizes and distances, he established that the Sun must be much larger than the Earth, whereas the Moon was smaller. This probably influenced him in advancing the heliocentric hypothesis because it must have seemed unreasonable that the large body (the Sun) should be revolving around the smaller one (the Earth).

For well over a thousand years after these first attempts at measuring the distances and sizes of the planetary system, rel-

atively little progress was made. Improved methods of measurement led to an upward revision of the figures by Hipparchus, who lived from 190 to 120 B.C. From the observatory
he built at Rhodes he made many important contributions to
astronomical development, notably the discovery of the precession of the equinoxes. The accuracy of his measurements
depended on his invention of a new form of diopter that was
greatly superior to the original diopter of Archimedes.[12] From
measurements with this instrument Hipparchus concluded
that the sizes and distances for the Sun and Moon obtained by
Aristarchus were far too small. Whereas Aristarchus found
the diameter of the Sun to be 6.75 times that of the Earth,
Hipparchus measured it to be 12.3 Earth diameters. Similarly
he increased the distance from the 180 Earth diameters of
Aristarchus to 1245.

Posidonius lectured at Rhodes during the first half of the
first century B.C. and must have been well acquainted with the
work of Hipparchus. Nevertheless, his further upward revision
of the sizes and distances seem to have been based on an idea
of Archimedes about the size of the solar orbit rather than on
more accurate measurements. Whatever his source,
Posidonius calculated the diameter of the Sun to be 39.25
Earth diameters and the distance to be 6550 diameters. Thus,
Posidonius obtained a value for the size of the Sun that was
approximately one third of the modern value and a distance
about one half of the correct value. It is a most salutory
thought that two thousand years ago astronomers knew the
dimensions of the planetary system far more accurately than
we know the distances to the remote objects in the universe on
which our modern cosmology is based.

Although these measurements and estimates did not depend on any belief in a geocentric or heliocentric hypothesis,
it is obvious that no incentive existed for radically new types of

12. Hipparchus mounted two vertical plates, one fixed and the other movable, on a
horizontal support. There was one hole in the fixed plate and two holes, one above the
other, in the movable plate. He viewed the Moon (for example) through the hole in
the fixed plate and then moved the other plate until the two holes coincided with the
upper and lower extremities of the lunar disk.

measurement until the acceptance of the heliocentric theory in the seventeenth century. In the meantime, during the one and half millennia separating Posidonius from Copernicus, estimates for the sizes and distances of the Sun showed an upward trend, apart from the curious regression by Ptolemy. Ptolemy did not believe in the calculations of Posidonius and thus reduced the distance estimate by ten times and the size estimate by eight times. Although Ptolemy introduced these erroneous values for the Sun, he gave nearly correct values for the Moon —a diameter of 0.29 terrestrial diameters and a distance of 29.5 terrestrial diameters (the correct figures are 0.27 and 30.2). Arabic astronomers too were well acquainted with the vastness of astronomical compared with terrestrial measurements. In the ninth century A.D. Al-Fargani calculated that the Moon was 64 1/6 Earth radii from the center of the universe—an overestimate by a factor of only 6% if the Earth was to be regarded as the center of the universe. In *De revolutionibus* Copernicus produced complex arguments for the relative distances and order of the various planets. He agreed with Ptolemy about the Moon, but his estimate of the distance of the Sun was far less accurate than that of Posidonius.

Although the heliocentric hypothesis became generally accepted in the first half of the seventeenth century, little was really known about the distances involved in the planetary system. There was good agreement about the distance of the Moon, but solar distance estimates varied by several times. Copernicus had settled on a value almost one tenth less than that derived by Posidonius seventeen hundred years earlier. In 1618, Kepler enunciated his third law of planetary motion and provided the foundation whereby this complex matter could be settled.

Kepler was unable to discover any parallax for Mars, and he recognized that this implied the need for increasing the accepted values for the distances. But in the *Epitome astronomiae Copernicanae* in which he published the third law there is a curious argument that leads him to conclude that the distance to the Sun must be three times greater than the currently accepted value—and that was still only one seventh of the true distance. The third law, relating the square of the period of a planet to the cube of the distance from the Sun, provided the basis for the first real advance in mensuration since the age of Aristarchus and Hipparchus.

The time taken by a planet or the Earth to complete an orbit of the Sun could be determined without difficulty; thus the measurement of a single distance, Sun to Earth or Sun to a planet, enabled the scale of the entire planetary system to be calculated. In essence the problem resolved itself into a direct measurement of the distance of the Sun from the Earth—and that had to be accomplished by measuring the parallactic angle from a baseline on Earth. Even in modern times, the direct measurement of the parallax of the Sun by straightforward geometrical methods from the maximum possible baseline on Earth is not the technique adopted. The angle, known as the solar parallax (i.e., the angle subtended at the Sun by the radius of the Earth), is extremely small, of the order 8.8 seconds of arc. This is equivalent to the angle subtended by a cent when viewed from a distance of 488 yards. Since the Sun is so large that it subtends an angle of half a degree at the Earth, direct measurement is evidently not a satisfactory procedure. But it was recognized that if the distance of one of the nearer planets—Mars, Mercury, or Venus—could be determined, then the application of the Keplerian third law would enable the scale of the entire planetary system to be settled.

The first notable advance came in 1671–73 when J. D. Cassini organized an expedition from the Paris Observatory to Cayenne to determine the shape of the Earth. At his suggestion Jean Richer made measurements of the angular distance of Mars from the vertical for comparison with the angular distances measured simultaneously by Flamsteed in England and Cassini in France. From these terrestrial baselines the parallax, and hence the distance of the planet, was calculated. Computations on the basis of Kepler's third law gave a value for the solar parallax of 9.5 seconds of arc. Although this represents a distance about 8 percent less than that subsequently established, it settled a major discrepancy in astronomical measurements that had existed since ancient times. These measurements resolved the extreme uncertainties about the size of the planetary system, but some disagreement remained about the precise value of the Earth–Sun distance. In terms of the radius of the Earth, which by the late seventeenth century had been well established, the uncertainties in the measured parallax of 9.5 seconds of arc gave minimum and maximum limits of 82 million and 87 million miles for the distance of the Sun.

The baseline provided by the extreme seasonal positions of the Earth in its orbit around the Sun is critical for all astronomical distance measurements beyond the solar system. The possibility of using this baseline for the measurement of the parallax of the stars, and hence for the determination of the distance scale for the universe, was evident as soon as the heliocentric theory became accepted. Another one and a half centuries were to elapse before anyone achieved success in attempts at these stellar measurements. In the meantime, the establishment of the correct value for the baseline (i.e., for the solar parallax), became a major topic in observational astronomy, but the precise value remained a subject of contention for nearly three hundred years after this first measurement in 1672.

The initial revision of the 1672 results will forever be connected with Edmond Halley. Reference has already been made to Halley's association with Newton and the publication of the *Principia* in 1687. The wide scope of Halley's interests and scientific work is indicated by the fact that four years later, in 1691, Halley published six papers in the *Philosophical Transactions of the Royal Society*. The topics dealt with in these papers ranged widely: the circulation of sea currents, the place where Julius Caesar landed in Britain, the flight of birds, the refraction of light, and deepsea diving.

In the third of these papers Halley considered the problem of measuring the Sun's distance. Because Venus approaches the Earth more closely than Mars, he proposed that measurements should be made of this planet. He did not suggest a repetition of the measurement of the zenithal distance of Venus, as had been done for Mars, but made the important proposal that the observation should be made when Venus could be observed in transit across the disk of the Sun. The idea that this observation should be used to measure the Sun's distance occurred to Halley in 1677 when he was in St. Helena making a catalog of the southern stars. There, with a telescope of focal length 24 feet, he had observed the passage of Mercury across the solar disk and found that he could time accurately the moment when the planet entered the disk and when it emerged. In his 1691 paper, and subsequently, he considered that observations should be carried out with Venus because it approached closer to the Earth and presented a larger disk than Mercury.

Although Venus passes between the Earth and the Sun every nineteen months, the actual occasions when the planet can be seen in transit across the disk of the Sun are far less frequent because the plane of the planetary orbit is inclined to that of the Earth by 3 degrees 23 minutes. The transits usually occur in pairs, separated by about 8 years, but the intervals between such transits are never less than 117 years.[13] The precise methods and calculations involved in making these observations were published by Halley in 1716 when he outlined a detailed program for the next transit, which would occur in 1761.[14] For that observation, British expeditions set out for St. Helena and Sumatra.

Charles Mason and Jeremiah Dixon never reached Sumatra. Their ship was in battle with a French ship in the English Channel and had to return to Portsmouth. Eventually Mason and Dixon observed the transit from the Cape of Good

13. A transit across the central meridian of the Sun occupies about eight hours. In 1639, fifty-two years before Halley published his paper, a transit of Venus had been observed in England by Horrocks and Crabtree, who seem to have suggested that the observation might be used as a basis for measuring the solar parallax.

14. An explanation of the method of deriving the Earth–Sun distance from the timing of the transits may be found in C. A. Ronan, *Edmond Halley: Genius in Eclipse* (New York, 1969; London, 1970), chap. 8. Full details of the expeditions of 1761 and 1769 are given in H. Wolf, *The Transits of Venus* (Princeton, 1959).

Hope. Expeditions from France set out to observe the transits from Siberia, Madagascar, Vienna, and southeastern India. Guillane Le Gentil, who intended to observe from the southeastern Indian position, never reached his goal because his ship was attacked by the British fleet. Altogether, the transit of 1761 was observed from 62 sites, and that of 1769 from 63 sites. Halley had died at the age of eighty-two in 1742, but the results of these expeditions justified his belief that the observation of transits would provide an accurate measurement of the solar parallax. The distance calculated was 95 million miles, but a later analysis of the measurements reduced this to 93 million miles, within a quarter of a percent of the value accepted today.

For the next pair of transits, in 1874 and 1882,[15] expensive expeditions were organized by the British and the Americans, but timings differed by ten seconds and at that stage other techniques were being used to determine the solar parallax. In 1877 the English astronomer D. Gill observed the relative displacement of a planet against the stars as seen from two points on the Earth's surface. Gill used various planets, first Mars, obtaining a value for the parallax of 8.78 seconds of arc, and later several minor planets to avoid the difficulties introduced by the size of the disk of the planet. By 1900, using the asteroid Eros,[16] the value had been refined to 8.790 seconds of arc. In the twentieth century more subtle methods were used for determining the parallax: for example, observation of perturbations caused by the Earth's attraction in the motion of Venus or Eros, or the determination of the constant of aberration by spectroscopic measurements, at half-yearly intervals, of the relative velocity between the Earth and a star. By the middle of the century a strange situation existed. When all the attempts to determine the parallax were reviewed, the two most accurate methods were believed to be correct to 1 part in 10,000, but their results differed by 1 part in 1000. That is, the

15. The next pair of transits will occur on June 8, 2004, and on June 6, 2012.
16. The asteroid Eros was discovered in 1898. It approaches closer to the Earth than any other planet.

fundamental unit of distance in astronomy was known only to an accuracy of one tenth of one percent.

The resolution of this small but disturbing uncertainty about the actual distances of the Sun and the planets suddenly became a matter of prime importance with the rapid development of space technology after the launching of the first Soviet Sputnik in 1957. Within a few years rockets were developed that made it possible to launch payloads to the Moon and the planets. The Soviet Union launched the first space probe to Venus on February 12, 1961. It is a remarkable fact that at that moment, the distance of the planet was not known with sufficient accuracy to make contact between space probe and planet certain.[17] The development of large radio telescopes and powerful transmitters had by that time made it possible to measure the distance of the planet by radar. If a radar pulse could be transmitted to Venus and observed after reflection, the time of the journey, and hence the distance to the planet, could be measured with precision. The exact value of the Earth–Sun distance and the solar parallax could then be derived from Kepler's third law. In 1961 three independent radar measurements from the United States, the USSR, and the UK achieved success. At last the value of the astronomical unit, the mean distance of the Earth from the Sun, was derived to an accuracy of 1 part in 30 million.

The precision of this final resolution of a long-drawn-out problem has been well evidenced in recent years by successful landings on Mars and Venus by several American and Soviet spacecraft, while the radar value of the mean Earth–Sun distance of 149,600,000 km is accepted by the International Astronomical Union as an agreed figure, equivalent to a solar parallax of 8.794 seconds of arc. There is some irony in the recollection that this precision radar measurement differed by 60,000 km from the best previously accepted values of the conventional optical and dynamical determinations—and these were claimed to be accurate to at least 15,000 km.

17. The Soviet scientists intended to use the transmitter controlling the probe to measure the planetary distance by radar during the flight of the probe to the planet. In this they were successful, but during the flight of the probe to the planet, contact with the spacecraft failed.

11

The Size of the Universe: The System of Stars and Galaxies

The first measurement of the distance to a star was not accomplished until 1838. Before that year, throughout the entire period of the development of astronomy, ideas of the size of the stellar system were highly speculative, although after the acceptance of the heliocentric theory certain limits were imposed on the minimum distance to the stars. Ancient cosmology was much concerned with the processes by which the motion of the sphere of the fixed stars was transmitted to move the Sun and the planets. In the cosmology of his *Epinomis* Plato introduced the idea of the space-filling ether. The concept of a sphere of ether removed the constraints on the need for a small universe, and so we are told in the *Epinomis* that the stars are "immense," although no distances are mentioned.

Aristotle based his cosmology on the concept that the entire universe was contained within the sphere of the stars, that it was finite, and that there was "neither place nor void" beyond this heaven. For Aristotle the highest heaven was a boundary beyond which no place existed. Although the protagonists of the infinity of space and the existence of a void regarded the Aristotelian idea as absurd, there seems little evidence of any realization of the dimensions of the stellar system in terms of terrestrial measurements. The Aristotelian concept of space-

filling shells transmitting movement from the sphere of the
fixed stars became an important link in defining ideas about
the size of the universe in the light of the Ptolemaic system.
Every sphere had to be large enough to contain each planet's
set of epicycles and circles. The relative sizes thus derived
could be transformed to actual distances on the basis of the
current estimate of the distance to the sphere of the Moon.
Estimates of the size of the universe made on this basis became
common, especially among Arabic astronomers. In the ninth
century Al-Fargani calculated in this manner the distance of
the sphere of the stars to be 75 million miles from the earth.[1]
Since Al-Fargani calculated the distance of the Moon to be
208,000 miles, it is evident that astronomers of the time were
well aware of the minuteness of the Earth and the sublunary
region compared with the dimensions of the universe of stars.

With the acceptance of the heliocentric hypothesis the first
real constraints were placed on the minimum distance to the
stars. *De revolutionibus* contains a complicated argument about
the consequences following the removal of the Earth from its
fixed central position in the universe. If the Earth is stationary
at the center of the sphere of fixed stars, then diametrically
opposite stars on the sphere will be observed to rise and set at
the same moment. On the other hand, if the Sun is the center
of the stellar system and the Earth is in motion around the
Sun, then the rising and setting of opposite points on the
stellar sphere cannot be simultaneous. Copernicus could only
set limits to the angle at which, for example, the summer
solstice was above the eastern horizon when the winter solstice
had just reached the western horizon. On these arguments the
Copernican system required the sphere of fixed stars to be at
least 1.5 million Earth radii distant. That was seventy-five
times greater than Al-Fargani's estimate and gave the distance of the stellar sphere as at least 5600 million miles.

The volume of the heliocentric universe of Copernicus had
to be vast compared with that of the ancient cosmologies, leav-

1. On the basis that the radius of the Earth was 3250 Roman miles.

ing an immense space between the most distant known planet, Saturn, and the sphere of the stars. This troubled Tycho Brahe. In his rejection of the heliocentric theory he emphasized that Copernicus had been forced to open up this vast space between Saturn and the stars merely in order to account for the absence of an observable parallax. Even with his superior instruments Tycho failed to measure any parallax. The limits he placed on this were far smaller than any previously assigned, and he argued that if the Earth was in motion, then his failure to detect stellar parallax could only imply that the stellar sphere was seven hundred times farther from Saturn than the distance between Saturn and the Sun. On contemporary estimates this would have placed the sphere of fixed stars at a distance greater than 40,000 million miles. Such distances of more than 10 million times the radius of the Earth, and the implication of vast regions of emptiness between stars and planets, were inconceivable even to Tycho.

Soon, with the general acceptance of the heliocentric theory, the true interpretation had to be placed on the failure of Tycho to measure any parallax for the stars, namely, that stars *were* immensely distant compared with the distance of Saturn, the farthest known planet. Throughout the next two centuries, with the steady improvement of astronomical instruments, repeated attempts were made to measure the parallax of a star. Jacques Cassini of the Paris Observatory attempted to measure that of the star Sirius, basing his choice on the belief that because Sirius was the brightest star, it was also the nearest. Cassini observed Sirius with a transit telescope as it crossed the meridian, and having performed this measurement with the Earth in various positions of its orbit around the Sun, he claimed that he had detected a shift in the position of Sirius against the background of fainter stars. In 1720 Halley subjected Cassini's measurements to a critical assessment and concluded that Cassini could not have measured any effect caused by the parallax of Sirius.

Ten years earlier, Halley reached another revolutionary conclusion about the stars. Throughout history no one appears to have questioned the idea of the sphere of "fixed" stars. Halley examined the position of the stars listed by Ptolemy and by Hipparchus and compared these with more recent positions determined by himself and others. He came to the conclusion that differences in the positions were beyond any reasonable errors of measurement and argued that the only explanation must be that stars were not "fixed" in space but were possessed of their own motions. Halley's discovery of the "proper motion" of stars was not pursued until the end of the century when Herschel made new measurements and calculated the direction of the motion of the Sun in space.

A few years after Halley's rejection of the Cassini measurements on Sirius, James Bradley, the third Astronomer Royal of England, thought he had succeeded in measuring the distance of the star γ-Draconis. Born in Gloucestershire in 1693, Bradley was an extremely able astronomer. He learned his skills when living with his uncle, James Pound, rector of Wanstead in Essex, a distinguished amateur astronomer of that day. Bradley graduated at Oxford University in 1717 and then held various church appointments. Four years later he was elected to succeed Halley as Savilian professor at Oxford, but until his uncle died in 1724 he continued to live and observe with him at Wanstead. He became Astronomer Royal in 1742 and then lived at Greenwich, but his famous observations on the star γ-Draconis were made at Wanstead. This star nearly passes through the zenith in that latitude, and in 1669 Robert Hooke had observed annual displacements in its position. Both he and John Flamsteed, the first Astronomer Royal, who in 1694 had detected a similar motion of the pole star, thought they were observing the parallactic displacement of the stars as the Earth moved in its orbit around the Sun.

Bradley's observations on γ-Draconis were made from 1725 to 1728. He discovered from the apparent position of the star that it described a small ellipse as he observed it throughout the year; but he was puzzled by the fact that the displacement was farthest north in September, and farthest south in March; if the displacement were due to the effect of parallax, these maximum shifts amounting to 40 seconds of arc should be in

June and December. The solution of this problem occurred to Bradley in September 1728 when he was on a pleasure boat on the Thames. Noticing that the pennant on the masthead changed direction with the course of the boat, although the wind remained constant in direction, he translated this into the movement of the Earth and saw the explanation of the anomalous elliptical path of the position of the star as the Earth moved around the Sun. Shortly afterward he announced his discovery of the aberration of light to the Royal Society.[2] The detailed analysis of these observations also led Bradley to the conclusion that the actual parallactic displacement of the star γ-Draconis must be less than 2 seconds of arc and probably not more than half a second of arc. This lower limit assigned by Bradley would have placed the star farther away than 37 million million miles.[3]

The complete failure to detect any annual parallax for the stars and the limits placed on the distance of γ-Draconis by Bradley did not deter Herschel, fifty years later. Herschel believed that with his superior instruments he could succeed. The attempt by Herschel to use stars that lay close together in the sky for the determination of parallax has been described.[4] His assumption that the fainter of the pair would be the more distant and therefore "fixed" compared with the brighter star he presumed to be nearer was erroneous. Although Herschel's attempt to measure parallax failed, we have seen that his observations led him to the discovery of double stars. Not for the first, nor for the last, did the attempt to measure the scale of the universe, although failing in its primary purpose, lead to new discoveries of major importance.[5]

At last, in 1838, Friedrich Bessel at Königsberg succeeded

---

2. Light travels at a finite speed. When the light from a star reaches an observer on Earth, the relative positions of star and Earth are no longer the same as when the light was emitted. The aberration angle is determined by the ratio of the Earth's orbital speed to the velocity of light.

3. The star γ-Draconis is now known to be at a distance of 36 parsecs (1 parsec $\doteq$ 30.86 x 10^{12} km, or 3.26 light years). The parallax of the star is 0.028 seconds of arc, or about 18 times smaller than the lower limit given by Bradley.

4. See chap. 9.

5. Bradley, aberration of light; Herschel, double star systems; and in 1960–63 the search for more distant radio galaxies led to the discovery of quasars.

in measuring the trigonometric parallax of a star. In twelve years Bessel determined the position of 75,000 stars with a precision hitherto unapproached. Then he asked J. von Fraunhofer of Munich to construct for him a special device known as a heliometer, in principle a large double-image micrometer. With this Bessel measured the position of the star 61 Cygni over a period of a year and found that its apparent position changed against the background of the faint stars. In 1838 he was able to announce the measurement of the first stellar parallax—only 0.30 seconds of arc—representing a distance of eleven light years. It was a remarkable achievement, made possible by the combination of the superb instrument built by Fraunhofer and Bessel's own observational skills.

Less than three centuries after the promulgation of the heliocentric hypothesis, an extension of the cosmological distance scale to an amount inconceivable in terrestrial terms had occurred. Copernicus could still envisage a minimum limit for the distance of the stellar sphere in terms of the size of the Earth (1.5 million Earth radii). Tycho's limit was seven hundred times farther than Saturn. Bradley's limit of 37 million million miles was already beyond human comparison, and this measurement of Bessel revealed that the star 61 Cygni was at twice this distance.[6]

The extraordinary atmosphere of those days is revealed in the archives of the Royal Astronomical Society. Two years after Bessel's achievement, the council of the Society was unconvinced that the attempts to measure the parallax had succeeded. In explaining why the Society's Gold Medal for 1840 was not being awarded to Bessel, President Sir John Herschel[7] said that the "observations which [are,] it would appear, beyond question, have brought us to the very threshold of that long-sought portal which is to open to us a measurable pathway into regions where the wings of fancy have hitherto been overborne by the weight or baffled by the vagueness of the illimitable and the infinite." But after describing Bessel's

6. The nearest stars known today are about 25 million million miles from the Sun.
7. Son of Sir William Herschel.

measurements as they were available to the Society, Sir John went on to say:

> It may now be reasonably asked, if all this be so, why have your Council hesitated to mark this grand discovery with that distinct stamp of their conviction and applause, which the award of their annual medal would confer? A problem of this difficulty and importance solved, so long the cynosure of every astronomer's wishes—the ultimate test of every observer's accuracy—the great landmark and *ne plus ultra* of our progress, thus at once rooted up and cast aside, as it were, by a *tour de force,* ought surely to have commanded all suffrages. It is understood, however, that we have not yet all M. Bessel's observations before us. There is a second series, equally unequivocal (as we are given to understand) in the tenor, and leading to almost exactly the same numerical value of the parallax, and not yet communicated to the public. Under these circumstances, it became the duty of your Council to suspend their decision. But, should the evidence finally placed before them at a future opportunity justify their coming to such a conclusion, it must not be doubted that they will seize with gladness the occasion to crown, with such laurels as they have it in their power to extend, the greatest triumph of modern practical astronomy.[8]

A year later Sir John and the council were satisfied with Bessel's measurements. In his address on 12 February 1841 announcing the award of the Gold Medal to Bessel, Sir John paid tribute to the instrument "with which Bessel made these most remarkable observations." Concluding his address, Sir John said, "Gentlemen of the Astronomical Society, I congratulate you and myself that we have lived to see the great and hitherto impassable barrier to our excursions into the sidereal universe; that barrier against which we have chafed so

8. *Mon. Not. Roy. Astr. Soc.* 5, no. 4 (Feb. 14, 1840): 32.

long and so vainly—(*aestuantes angusto limite mundi*)—almost simultaneously overleaped at three different points. It is the
greatest and most glorious triumph which practical
astronomy has ever witnessed."[9]

Although Bessel's measurement was rightly assessed by Sir
John Herschel as a "glorious triumph," it is doubtful if
astronomers of that day had any appreciation of the great difficulty of extending these parallax measurements to other
stars. The attempt to grasp the immensity of the stellar system
revealed by Bessel's measurement obscured the reflection that
even these distances of tens of millions of millions of miles
might be insignificant compared with the extent of the stellar
system visible to the unaided eye. Yet by the end of the nineteenth century astronomers had succeeded in measuring the
parallax of only another seventy stars, and some of these were
listed as doubtful. Although photographic techniques were
proving of great assistance in these measurements, an
astronomer speaking at one of the Royal Institution Discourses in London in 1908 said that "with the exception of a
hundred stars at most, we know nothing of the distances of the
individual stars."[10] The further development of telescopes and
techniques was soon to change this situation, but it may be
remarked that even with the best contemporary techniques of
measurement the reliability of parallax measurements has
been improved by only about 50 times over that achieved by
Bessel in 1838. Modern star catalogs contain direct
trigonometrical parallax measurements on stars at distances
of about 100 parsecs (approximately 2000 million million
miles). Within this range there are some 10,000 stars.

The realization that the direct measurement of parallax was
likely to be confined only to the nearer stars provided a great
stimulus to discover other methods by which the distance of
the remote stars could be determined. An extension of the par-

9. *Mon. Not. R. Astr. Soc.* 5, no. 12 (Feb. 12, 1841): 89. The reference to "three different points" was to similar measurements then being announced by Thomas Hen-
derson and W. Struve.
10. Jacobus Kapteyn, on May 22, 1908. The lecture has been reprinted in the *Royal
Institution Library of Science: Astronomy* (Netherlands; Elsevier, 1970), 2:78.

allax method, using a baseline provided by the motion of the Sun through space, was developed. An observer on Earth is translated through space by 20 km every second because of the motion of the entire solar system. This provides a baseline that continually increases and is equivalent in one year to twice the diameter of the Earth's orbit around the Sun. This method—known as secular parallax—has to take account of the proper motion of the star itself through space and is most useful when large groups of stars are involved so that the group as a whole may be taken to be at rest. This and other spectrographic methods extended the distance measurement of stars well beyond 100 parsecs, but the major advance that was to lead to our present understanding of the extent of the Milky Way and the real nature of nebulae occurred in 1908.

Henrietta Leavitt of the Harvard Observatory had studied certain types of stars in the Small Magellanic Cloud. These stars are known as Cepheid variables, a class named after the star δ-Cephei, which changes its brightness by a magnitude, from maximum to minimum, in five days. She noted the curious fact that in the case of the brighter stars, the period of light variation from maximum to minimum was longer and, indeed, she found that there was a definite relationship between the brightness of the star and its period of light variation. There were Cepheid variables among stars whose distance could be estimated from spectrographic methods. Harlow Shapley, working with the 60-inch telescope at Mt. Wilson, calibrated Henrietta Leavitt's apparent-brightness/period relationship in terms of absolute magnitude. It was then possible, by measuring the period of a Cepheid variable, to find its absolute magnitude, and hence, by observing its apparent magnitude, to determine its distance by the inverse square law.

The Size and Structure of the Milky Way

We have seen that although Herschel concluded that the

stars of the Milky Way were not distributed symmetrically around the Sun, and that he established an approximate shape for the system, he could make no further progress in the absence of any means of measuring stellar distances. His work had destroyed the ancient and long-held belief in the sphere of fixed stars (although Halley had demonstrated that they were not "fixed"), but there was no real appreciation of the position of the Earth and solar system within this star system. Consequently, the general belief that the Earth must be situated at or near the central region of the Milky Way remained. This age-old egocentric concept of man's place in the universe was finally eradicated soon after Henrietta Leavitt's discovery of the period-luminosity relationship for the Cepheid variable stars.

By 1918 Harlow Shapley had used this method to estimate the distance of twenty-five globular clusters. About a hundred of these globular clusters are known. They are nearly spherical groups of more than a hundred thousand stars concentrated toward the band of the Milky Way. Although nearly all these globular clusters had been cataloged by Herschel, Shapley seems to have been the first astronomer to consider the implication of their distribution in the heavens. Assuming that the clusters defined the general shape of the Milky Way, and having the means to determine their distance, he was able to outline the nature of the Milky Way as we understand it today, namely, as a system of 100 billion stars distributed throughout a flattened disk 100,000 light years across. His discovery that the Sun was situated in a spiral arm of this disk, 33,000 light years from the central region, finally eroded the belief that on Earth man was placed in a central and favored position among this system of stars.

Shapley published his conclusions in 1918 and 1919 in the *Astrophysical Journal* and in the *Contributions* from the Mt. Wilson Observatory. His results were not received without opposition. On April 26, 1920, at a famous meeting of the National Academy of Sciences in Washington, Shapley defended his conclusions. He said that "the Sun is found to be very distant from the centre of the galaxy. It appears that we are near the centre of a large local cluster or cloud of stars, but that cloud is at least sixty thousand light years from the galactic centre . . . we have been victimised by the chance position of the Sun near the centre of a subordinate system, and misled by the consequent phenomena, to think that we are God's own appointed, right in the thick of things."[11]

It will be noticed that Shapley gave the distance of the Sun from the galactic center as "at least 60,000 light years"; this is about twice the figure of 33,000 light years accepted today. Shapley must have been the first astronomer in history to overestimate an astronomical distance. The various discrepancies that prevailed in the estimates of the size of the Milky Way were eventually resolved in 1930 when R. J. Trumpler of Lick Observatory published his results on the interstellar absorption of starlight. Shapley had neglected the effect of such absorption and had therefore made erroneous estimates of the apparent magnitude of the Cepheids in the globular clusters. Although Shapley's conclusions were disputed when they were announced, and this uncertainty about dimensions continued to exist for a decade, within a few years it would probably have been difficult to find an astronomer who still defended the ancient and tenaciously held idea that we are near the center of the star system. It is curious that in spite of his great accomplishments, Herschel failed to reach this conclusion more than a century earlier. Although he could not have placed a distance scale against the star system, he had plotted the positions of most of the globular clusters including all those measured by Shapley. It was Shapley's great achievement to recognize the significance of their distribution in relation to the position of the Sun in the star system.

Distances of the Nebulae

We have seen that Herschel cataloged 2500 nebulae. His

11. The original typescript of this address is in the archives at Harvard. It was reprinted in *Jr. Hist. of Astr.* 7 (1976): 175.

ability to resolve some of these objects into stars by the use of his large telescope led him to conclude that nebulae must be star systems existing in their own right, well outside the Milky Way. It is questionable whether Herschel maintained this confidence to the end of his life, for he had been unable to resolve some of the nebulae into stars. And at the beginning of the present century it is probably true that most astronomers believed the nebulae to be wholly contained within the Milky Way system, and that this represented the totality of the observable universe. However, as the discovery of the Cepheid variable star relationship provided a means of measuring distance that enabled Shapley to settle the problem of the Milky Way, so the same technique soon gave the definitive answer to the problem of the nebulae.

Even before those direct measurements were made, a body of astronomical opinion reached the conclusion that Herschel was correct in believing that some nebulae were extragalactic star systems. For example, Heber Curtis of Lick Observatory compared the maximum magnitudes of five novae in four spiral nebulae with those of novae in the Milky Way and concluded that these spiral nebulae were distant from the Milky Way (about 20 million light years). Reference has been made to the debate in Washington in 1920 when Shapley defended his conclusions about the shape and size of the Milky Way. Curtis was one of his opponents at that time, but curiously Shapley believed that Curtis was wrong in his conclusion that the spiral nebulae were distant objects far from the Milky Way. Although Shapley was right about the Milky Way and Curtis was right about the nebulae, both men overestimated the distance involved (some novae used by Curtis were actually supernovae), which helped confuse the issue.

Within a few years of that Washington debate, Edwin Hubble, using the new 100-inch telescope on Mt. Wilson, was able to settle the argument decisively. In 1926 he published his results on the nebula M33. He had been able to resolve the nebula into stars, among which were Cepheid variables. Then, as Shapley had done for the Milky Way, Hubble was able to measure the distance of M33 by using the period-luminosity relationship for the Cepheids. Although Hubble's results were conclusive to the extent that the nebulae in which he could resolve stars were star systems in their own right, existing in space beyond the confines of the Milky Way, he underestimated the distances. At that time it was not realized that there were two populations of Cepheids and that they differed in their period-luminosity relationships. Thus Hubble underestimated the distances of the nebulae by approximately a factor of 2. For example, his distance estimate for M33 was 720,000 light years, whereas today we estimate the distance of that nebula to be 2,360,000 light years.[12]

We now realize that Messier's and Herschel's catalogs listed four different types of objects as nebulae, and that three of these types are in the Milky Way. Therefore it seems probable that the argument about the nature of nebulae would have continued for many years had not the new 100-inch telescope had just sufficient light-gathering power to detect the Cepheids in a few of the nearer nebulae. This enabled Hubble to make a statistical study of four hundred nebulae on the assumption that those of the same type would have the same absolute magnitude. Thus, for all nebulae of the same spiral formation as M31, for example, in which he had measured the Cepheids, he deduced their distances by comparison with the luminosity of M31. He reached the conclusion in 1926 that the 100-inch telescope could penetrate into space to a distance of 140 million light years and that this volume of space contained 2 million extragalactic nebulae.

As soon as Hubble established the extragalactic nature of the spiral nebulae, it became clear that the objects listed in the catalog of Herschel as nebulae covered a number of different categories. Three were within the local galaxy or Milky Way system. Typical of these were the nebulous clouds of gas like the Orion nebula, the remnants of supernova like the Crab

12. See E. Hubble, *The Realm of the Nebulae* (New York, 1958), p. 143, for Hubble's value; for the present value, see the paper by G. de Vaucouleurs *Astrophys. J.* 223 (1978): 730.

Nebula, and the "planetary" nebulae that often contain a central star. The reason for the uncertainties existing for centuries became evident. Of those cataloged by Hubble as distant extragalactic systems, there were various formations. The majority (about 80 percent) showed a spiral formation with the majority of the stars contained in arms radiating from a dense central core. Apart from a few of irregular formation, the remainder were spherical or elliptical galaxies exhibiting little structure. For more than twenty years it was believed that galaxies of these types represented the total content of the universe, and an evolutionary sequence from the spherical galaxies to the spirals was proposed by Hubble. Then, with the development of radio astronomy after World War II, other remote objects were discovered: in 1951, the radio galaxies and a decade later, the quasars. Their true nature, their place in the evolutionary sequence, and the processes by which such great energies are emitted from regions of space that are small by the standard of normal galaxies are not yet understood, but their discovery has extended our penetration into the universe to immensely greater distances than envisaged by Hubble in 1926. The introduction into astronomical research of the 200-inch Hale telescope on Palomar in 1949 extended the penetration to about 2 billion light years, and today the most remote quasars have been identified at distances of probably more than 6 billion light years.

In these studies of the extragalactic universe the problem of distance measurement still presents a severe difficulty. Cepheid variables can be measured in galaxies to a distance of a few million light years only. Within that range the distances obtained from these measurements may be compared with results obtained by a number of other methods, for example, with the distances derived from measurements of the magnitude of globular clusters, novae, and supernovae. These various methods give distances in reasonable agreement out to a few million light years. In the regions of space that lie beyond a few million light years, the estimates of distance become increasingly uncertain and are constantly open to revision and discussion. Based on the assumption that recognizable types of objects (e.g., supernovae or globular clusters) in the distant galaxies are similar to objects of the same type in nearby galaxies, measurements of distance can be extended to about 150 million light years. Beyond that, the similarity of whole galaxies has to be considered (the statistical method used by Hubble in 1926), and eventually the comparison of entire clusters of galaxies extends the distance scale to some billion light years.

The Cosmological Distance Scale

Although the possibility that star systems existed in space beyond the Milky Way had been debated for centuries, no one considered that the universe might exhibit phenomena of large-scale movement. The orbital motion of the Earth and planets around the Sun, the motion of the Sun and planetary system through space, and the proper motion of the stars are at high velocities compared with motions on Earth. But these velocities are small compared with the velocity of light. In 1929, a few years after the identification of the Cepheids in certain nebulae, Hubble published evidence that there was a cosmic expansion of the universe at very high velocity and that this increased linearly with the distance.

The observational foundation for these conclusions had been laid earlier by V. M. Slipher at the Lowell Observatory, Flagstaff, Arizona. In 1912 he obtained a spectrum of the M31 spiral nebula and found that the spectral lines, which could be identified from certain elements, were not in their correct position on the spectograph. The shift was toward the violet end of the spectrum. The obvious cause of such a shift in wavelength is that the source is in motion with respect to the observer— the Doppler effect—and a shift toward shorter wavelengths indicates that the source and observer are approaching each other. Slipher calculated that the velocity of approach was 190 miles per second.

It soon seemed that the result on M31 was unusual in that
the majority of measurements showed that the spectral lines
were displaced toward the red, implying increasing separation
of source and observer. By 1929 forty-six spectral measurements had been made. Among these were distance measurements for eighteen isolated nebulae and for the Virgo
cluster. On this basis Hubble found the relationship between
the displacement of the spectral lines of a galaxy to the red
and the distance of the galaxy: the apparent velocity of recession of a galaxy was linearly related to its distance from the
observer.[13] If the shift in the spectral lines to longer
wavelengths was interpreted as a Doppler effect, the implication was that in this sample of galaxies, velocities of recession
up to 1200 miles per second were involved.

In the succeeding years Hubble and M. L. Humason continued these measurements, using the 100-inch Mt. Wilson
telescope. In 1931 they published more results, which extended the redshift measurements to galaxies at a distance of
about 100 million light years, where the implied recessional
velocities were 2000 miles per second. Four years later these
data extended to distances estimated to be 230 million to 240
million light years, where the redshift measurements indicated
velocities of 26,000 miles per second—a seventh of the velocity
of light. The linear law connecting the distance and the
redshift still held. Hubble was cautious about the meaning of
this redshift; that is, he was cautious about the extent to which
actual velocities of motion of the galaxies through space were
involved. In his Silliman memorial lectures at Yale in 1935 he
said: "Because the telescopic resources are not yet exhausted,
judgment may be suspended until it is known from observations whether or not redshifts do actually represent motion.
Meanwhile, redshifts may be expressed on a scale of velocities
as a matter of convenience. They behave as velocity-shifts behave and they are very simply represented on the same familiar scale, regardless of the ultimate interpretation."[14]

13. E. Hubble, in *Proc. Nat. Acad. Sci. Amer.* 15 (1929): 168.
14. E. Hubble, *The Realm of the Nebulae, pp. 122-23.*

The commissioning of the 200-inch Hale telescope in 1949, with its greatly increased light-gathering power, led to redshift measurements of extragalactic nebulae estimated to be at distances of a few billion light years. The discovery of radio galaxies gave rise to even deeper penetrations into space, and in 1960 R. Minkowski used the 200-inch telescope to photograph a cluster of galaxies in the vicinity of one of these strong radio-emitting objects in the constellation of Boötes. From the apparent magnitude of the galaxies, the distance was estimated to be 4.5 billion light years. The shift in a spectral line, identified as an oxygen line, implied a recessional velocity of 86,000 miles per second, or 46 percent of the velocity of light. The distances, estimated on the basis of the apparent magnitudes of galaxies, also implied that the linear relationship between distance and velocity of recession was maintained over this vast distance.

Out to the most distant radio galaxy the extension of the distance scale has been based successively on the recognition of distant objects, similar to nearby objects, whose distance has been established by other means. No such possibility exists in the case of quasars because, at least to the present day, they appear as a new class of object not conspicuously related to other components of the universe whose distances are known by other methods. The distance scale for quasars has therefore been wholly based on the assumption that the redshift is a cosmological effect associated with the expansion of the universe. Whether the Hubble linear relationship is maintained for such great redshifts is at present a matter of dispute among astronomers.

The distances calculated from this cosmological effect are also widely disputed because of varying estimates of the value of the Hubble constant (i.e., the relation between the redshift and the distance). The value of the constant estimated by Hubble was 530 km per sec per megaparsec.[15] The revision of the distance scale for the nearer galaxies, and extensions of the

15. A megaparsec is a million parsecs, or 3.262 million light years. It should be noted that although these velocities are high on the cosmic distance scale, when reduced to the nearer stars they are very small—about 50 cm per sec.

measurements, showed that this value was five times too large. At present the astronomical literature contains arguments in favor of values of the constant lying anywhere between 50 and 100 km per sec per megaparsec.

It is perhaps salutary to recall a comment made by Allan Sandage, one of an eminent group of astronomers deeply involved in these measurements over the last few decades, in a lecture in 1971. In claiming an accuracy of 15 percent for new measurements he was presenting, giving a value of 55 km per sec per megaparsec, he said: "It is sobering and is the better part of prudence to recall that but twenty years ago the value of the Hubble constant was thought to be 530 km per sec per megaparsec, also with a stated accuracy of 15 percent."[16] As to the greatest distance to which the modern telescopes penetrate into space, perhaps the only answer that will stand the test of time is many billions of light years. Two thousand years ago, Posidonius calculated the distance of the Sun from the Earth. His estimate was about one half the correct value. It seems unlikely that we know the distances of many of the objects that can be studied with modern telescopes to that order of correctness. Indeed, observations with contemporary astronomical instruments transfer us to regions of the universe where the concept of distance loses meaning and significance, and the extent to which these penetrations reveal the past history of the universe becomes a matter of more sublime importance.

16. A Sandage, in *Q. J. R. Astr. Soc.* 13 (1972): 282.

12

Space and the Universe of Einstein

Since ancient times man has disputed about the nature of space and the existence of a void. In the first half of the fifth century B.C. Parmenides developed a form of metaphysically logical argument that, when applied to the problem of the existence of a void, led to much philosophical discussion. Among the doctrines set forth by Parmenides in *On Nature* is this somewhat obscure argument: "The thing that can be thought and that for the sake of which the thought exists is the same; for you cannot find thought without something that is, as to which it is uttered." This has been interpreted to imply that there can be no thought corresponding to a name unless the name is of something real; this leads directly to the view that there can be no change, because change means that things come into being or cease to be. Parmenides asserts: "Thou canst not know what is not—that is impossible—nor utter it: for it is the same thing that can be thought and that can be."

Leucippus, who flourished about 430 or 440 B.C. tried to reconcile the arguments of Parmenides with the clear evidence that motion and change did exist, and he conceded that there could be no motion without a void. It seemed that one must either agree that the world was unchanging or admit the existence of the void. The argument of Parmenides that a void

could not exist seemed logically irrefutable, and this belief was
encouraged when it was discovered that whenever there appeared to be a void, there was in fact air. Indeed, Parmenides
presented the ancient philosophers with a problem of extreme
difficulty because if one asserts that the void exists, the logical
response is that if the void exists, then it is not nothing and
therefore is not the void.

The escape from this dilemma was discovered by Aristotle,
who emphasized the distinction between space and matter. In
his *Physics* he stated that "the theory that the void exists involves the existence of place: for one would define void as place
bereft of body." The opinion of Aristotle that the distinction
between space and matter was important and that space was
a receptacle in which matter might or might not exist became
a long held and common-sense view. Unlike major parts of his
physics and astronomy, this view survived the physical and
cosmological revolution of the Copernican-Galilean epoch.
This fundamental view of Aristotle was enshrined in the Newtonian theory, for without ambiguity Newton implied the existence of absolute space and of absolute motion in that space.

There was, however, a vital distinction between the large-
scale cosmologies of Aristotle and Newton. Aristotle denied
the existence of infinitely large phenomena on the grounds
that the universe was finite and that nothing could exist
beyond it. By contrast, Newtonian mathematics, defining the
inverse square law of gravitational attraction between two
bodies, led to the need for an infinite extension of the universe
in order to preserve its stability. (Newton's clear statement of
this fact in his letter to Bentley in 1682 has been quoted in
chapter 9.) The Newtonian universe was infinite and the properties of a body were independent of the absolute space in
which it existed. Further, Newton was clear that matter and
space were created by God and that He had initiated the motions of the particles in the beginning. Newton's clear
statement of these views was made in his *Opticks:*

All these things being consider'd, it seems probable to me,

that God in the Beginning, form'd Matter in solid, massy, hard, impenetrable, moveable Particles, of such Sizes and Figures, and with Such other Properties and in such Proportion to Space, as most conduced to the End for which he form'd them. . . . It seems to me farther, that these Particles had not only a *Vis inertiae* accompanied with such passive Laws of Motion as naturally results from that Force, but also that they are moved by certain active Principles, such as is that of Gravity.[1]

Newton's universe seemed logically complete. The extension of the universe to infinity, without the possibility of defining the distance of even the nearer stars, seemed a reasonable common-sense view. Especially powerful was the statement that an infinite space existed to contain the infinite number of bodies in the universe. In the *Principia* Newton wrote that space "in its own nature, without regard to anything external, remains always similar and immovable," and this view and the cosmology based on it prevailed for two and a half centuries.

Although Newton's opinions on the absolute nature of space and motion prevailed, there were important opposing arguments. Leibniz, in particular, held contrary views, and in a sense he presaged a concept of space that was to emerge over two centuries later with Einstein. In his *Monadology* and in *Principles of Nature and of Grace* Leibniz argues that space has no reality. The real counterpart of space lies in the arrangement of the *monads*. Leibniz did not believe that the universe contained a limited number of substances possessed of a conventional "extension" in space. He believed that substance was composed of an infinite number of monads. A monad had some of the attributes of a physical point and was endowed with nonphysical attributes. In any event, Leibniz regarded "space" as existing only in the three-dimensional arrangement of the monads. Later, Kant, in *The Critique of Pure Reason*,

---

1. *Opticks* (London, 1717). This passage occurs on pages 375–76 of the 2nd ed. It does not appear in the 1st ed. (1704).

evolved a doctrine of space and time that again abandoned any idea of an absolute space existing in its own right in favor of a metaphysical and transcendental concept.

Early in the eighteenth century an important statement of the complete subjectivist view had been made by Berkeley. In *The Dialogues* (1713) and in *De motu* (1721) Berkeley produced a far-reaching rejection of Newton's objective attitude. He wrote that "every place is relative, every motion relative. If all bodies are destroyed we shall be left with mere nothing, for all the attributes assigned to empty space are immediately seen to be privative or negative except its extension. But this when space is literally empty, cannot be described or measured and so it too is effectively nothing." In this rejection of Newton's ideas on absolute motion, Berkeley considered Newton's explanation of the motion of a fluid in a suspended cylinder. If a cylinder containing a fluid is suspended, one can make the cylinder and then the fluid rotate. If the cylinder is then held firmly, the fluid continues to rotate inside the cylinder until it is brought to rest by friction. "But if the outward cylinder be forcibly held fast, it will make an effort to retard the motion of the fluid; and unless the inward cylinder preserve that motion by means of some external force impressed thereon, it will make it cease by degrees."[2] Newton used this as an example of real, or absolute, rotation. Berkeley used this illustration to arrive at the opposite conclusion: that the role of the cylinder was simply to hold the fluid and provide a means of initiating the rotation. He believed that the rotation of the whole stellar system relative to the fluid in the cylinder was involved in the motion of the fluid and that the concept of absolute motion was invalid.

Notwithstanding these philosophical attacks on Newton's position, no cosmological or mathematical development challenged his laws of motion and gravitation until the early years of the twentieth century when Einstein developed the special and general theories of relativity. There was, indeed, disquiet

---

2. *Principia*, sec. 9 on "The circular motion of fluids," prop. 51, theor. 39, cor. 6.

about the concept of the absolute parameters in Newtonian theory. Curiously, there seems to have been less concern about the problems raised by the need for an infinite universe in Newtonian cosmology. In his Tarner lectures in 1938, when the problems of space had been revolutionized by the Einstein general theory, Eddington remarked that "we usually regard infinite Euclidean space as the simplest kind of space to conceive. One would have thought that the infinitude would be rather a serious obstacle to conception; but most people manage to persuade themselves that they have overcome the difficulty, and even profess themselves utterly unable to conceive a space without infinitude."[3]

It was not the difficulty of the infinitudes in Newtonian cosmology, but the problems of absolute space, absolute time, and absolute motion that eventually led to the rejection of that cosmology. When Einstein was a student in Zurich (he began his four-year course in October 1896), a paper from the French mathematician Henri Poincaré was read at the International Congress of Mathematicians in that city. In this paper Poincaré said: "Absolute space, absolute time, even Euclidian geometry, are not conditions to be imposed on mechanics; one can express the fact connecting them in terms of non-Euclidean space." At about this time Ernst Mach's *Science of Mechanics* was published. Mach was born in Moravia in 1838 and after various appointments in Graz and Prague became professor of physics at Vienna. Although today Mach's name is known widely because of the Mach number relevant to high-speed flight, his work had a far more profound effect on the development of philosophy and science. He was concerned with the physiology and psychology of the senses and maintained that physical phenomena could be explicable only on the basis of data perceived by the senses. In *Science of Mechanics* Mach discussed the Newtonian laws and held that they contained no self-evident principle—specifically, that absolute space and absolute time were meaningless because they were

3. A. S. Eddington, *The Philosophy of Physical Science* (Cambridge, England: Cambridge University Press, 1939), chap. 9.

impossible to define in terms of observation. In a further elaboration of the ideas enunciated by Berkeley 150 years earlier, Mach emphasized that a body could never be accelerated in any absolute sense but that the acceleration must always be relative to the stars: "When we say that a body preserves unchanged its direction and velocity *in space*, our assertion is nothing more or less than an abbreviated reference to the *entire universe.*"

The whole fabric of thought about these issues was revolutionized in two papers published by Einstein. The first, "On the electrodynamics of moving bodies," appeared in 1905 in *Annalen der Physik*.[4] This concerned the problem of uniform or unaccelerated motion." In the *Principia* Newton wrote that "absolute, true, and mathematical time, of itself and from its own nature flows equably, without relation to anything external,"[5] and of space that it was "absolute space, in its own nature, without relation to anything external, remains always similar and immovable."[6] The essence of Einstein's 1905 paper was that space and time are relatively different in moving and stationary systems. A clock in motion runs more slowly with respect to one that is stationary; a measuring rod in motion shrinks with respect to one that is stationary; the velocity of light *in vacuo* is constant and is the limiting velocity in the universe. The many features that led Einstein to these conclusions, including the null result of the Michelson-Morley experiment and the various consequences of the special theory (including the equivalence of mass and energy) have been the subject of a vast literature in this century.[7] On the question of the *reality* of the effects predicted by the special theory, the material world has exhibited compulsive evidence. In matters of pure mensuration the quantity remains indeterminate until

4. The third of three historic papers published by Einstein in that year. The first applied the quantum theory to the photoelectric effect (for which Einstein was awarded the Nobel Prize in 1921) and the second concerned a mathematical analysis of the Brownian motion in liquids.
5. *Principia*, definitions, scholium 1 (Cajori revision of the 1729 Motte translation).
6. *Principia*, definitions, scholium 2.
7. For a comprehensive general account, see, for example, Ronald W. Clark, *Einstein. The Life and Times* (New York and London, 1973).

the observer is specified. "Those who associate with the result a mental picture of some entity disporting itself in a metaphysical realm of existence do so at their own risk; physics can accept no responsibility for this embellishment."[8]

The extension of the special theory concerned with uniform motion to the problem of bodies in accelerated motion occupied Einstein from 1907 to 1915. A series of publications during those years culminated in "The Foundations of the General Theory of Relativity" published in the *Annalen der Physik* in 1916. The problem of the distinction between the gravitational mass and the inertial mass of a body had existed since the enunciation of the Newtonian laws. The gravitational mass of a body is that to which we refer when calculating the force of attraction between two bodies. In his second law of motion Newton stated that the acceleration of a body depends on the force applied to it and on another property, which he called the "quantity of matter," possessed by the body—that is, the inertial mass—the inertia of the body being its resistance to motion. As Newton formulated his laws the gravitational mass and the inertial mass of a body are distinct entities. Further, since the gravitational mass determines the force exerted on another body, whereas the inertial mass is a measure of the resistance of the body to motion, the functions of these two properties of a body appear to be entirely different. The problem arose because neither Newton nor any of his successors was able to measure or detect by experiment any difference between these two quantities. (In 1840, for example, Eötvös proved that the inertial and gravitational masses of a body were equal to 1 part in 100 million.)

Further illustrations of the problem arise in two different methods by which the speed of rotation of the Earth can be determined. Without reference to the environment external to the Earth, one can make a dynamical determination by using a gyroscope or a Foucault pendulum. Alternatively, by observing a distant star, one can also measure the speed of rotation

8. Eddington, *The Philosophy of Physical Science*, chap. 5.

of the Earth. The two methods give the same result. Such coincidences found expression in Mach's principle that the local inertial frame must be determined by some average of the motion of distant objects in the universe. This implies that whereas the gravitational effect of a body is determined solely by its own properties, the magnitude of the inertia of the body is determined by the masses in the universe and by their distribution. If Mach's principle is correct, then the ratio of the two effects (the constant of gravitation) must contain information about the entire universe. This feature endows the general theory of relativity with vital importance to cosmology. Indeed, in the development of the cosmological implications of the general theory, Einstein wrote: "In a consequential theory of relativity there can be no inertia of matter against space but only inertia of matter against matter. If therefore a body is removed sufficiently far from all other masses of the universe its inertia must be reduced to zero."[9]

In the mathematics of the general theory Einstein sought to unify the gravitational and inertial fields in order to give expression to Mach's principle. To the extent that the equations imply an influence of massive bodies on the inertia of other bodies, he was successful. Nevertheless, difficulties of reconciliation with Mach's principle arose because the equations expressed the influence but not the entire cause of inertia, since the boundary conditions at infinity were not specified. As Einstein found it impossible to choose boundary conditions so that the inertial effects were fully determined, he introduced a constant (the cosmological constant) multiplying the inertial field terms in the equation. This cosmological constant had dimensions of the inverse square of length. With a positive value of this cosmological constant, Einstein was able to find a solution to the equations specifying a universe with uniform density of matter, random velocities zero, and with space so curved that it was unbounded but finite. Thus the difficulties at infinity were abolished, and since with the positive cos-

9. A. Einstein, *Sitzungsberichte der Preussischen Akademie der Wissenschaften* (Berlin, 1917) p. 142.

mological constant he believed there was no solution of the equations for empty space, he was of the opinion that Mach's principle had been fully incorporated in the theory. At that time all measured velocities were believed to be small compared with the velocity of light, and the universe was thought to be a static entity. Einstein concluded, therefore, that his equations represented the universe to a first approximation.

The subsequent development of theoretical and observational cosmology during the next few years presents a most remarkable feature of astronomical history. Within months of the publication of Einstein's 1917 paper, Willem de Sitter of Leiden showed that there was another solution of the Einstein equations for an empty universe: "Einstein's solution of the equations implies the existence of a 'world matter' which fills the whole universe. . . . It is, however, also possible to satisfy the equations without this hypothetical world matter."[10] Further, de Sitter showed that this universe would be static only if empty of matter, and that hypothetical test particles introduced into the universe would recede from each other with ever increasing velocity. De Sitter wrote that "all these results sound very strange and paradoxical" and "we can then say that all the paradoxical phenomena (or rather the negation of phenomena) which have been enunciated above can only happen after the end or before the beginning of eternity." In spite of the strangeness of de Sitter's conclusions, he demonstrated the important fact that Einstein was incorrect in his belief that by introducing the cosmological constant, he had discovered a unique model for the universe that incorporated Mach's principle.

Within a few years of the publication of de Sitter's paper, observational evidence emerging from the measurements of Slipher and Hubble conflicted with the concept of a static universe. (see chapter 11) It was a development that stimulated the search for theoretical solutions of the equations of general relativity intermediate between the static universe of Einstein,

10. W. de Sitter, "On Einstein's Theory of Gravitation and Its Astronomical Consequences," *Mon. Not. R. Astr. Soc.* 78 (1917): 3.

containing matter, and the empty universe of de Sitter, exhibiting motion without matter. It was claimed that, in any case, the density of matter in the universe must be very low and hence that solutions should exist close to that found by de Sitter. The solutions were found four years later by the Russian mathematician A. Friedmann, who showed that there was a whole family of cosmological models for the universe based on general relativity, in which the mean density of matter in the universe varied with time. Many theorists studied and developed the various possible cosmologies based on Einstein's general theory, and in those years when Hubble's observations were establishing the observational basis for the expanding universe, the theoretical work of G. Lemaître, A. S. Eddington, and H. P. Robertson was of great significance.

The various world models derived from the general relativistic equations have the characteristic of a universe evolving so that the radius increases and the mean density decreases as time increases. All pose the problem that a singular condition existed at time zero with the physical implication of a beginning of infinite density. The subsequent behavior depends on the assumptions made about the constants in the equations. If the mean density exceeds a critical value, then eventually the gravitational forces will overcome the forces of expansion and the universe will collapse. If not, the expansion of the universe will continue indefinitely, either monotonically or at a rate tending to zero as the radius tends to infinity. In principle, the actual observation of the universe should enable a distinction to be made between the various theoretical world models embraced in the relativistic theories. A feature of modern cosmology is the continuous and indecisive discussion about these issues.

13

The Age of the Universe

When Hubble delivered the Silliman lectures at Yale University in 1935, he estimated that on the average the universe contained one extragalactic nebula for every 5×10^{18} cubic light years of space. Taking the limiting magnitude that could be recorded by the 100-inch Mt. Wilson telescope, he considered it possible to penetrate 500 million light years into space and said that "about 100 million nebulae may be expected within a sphere of this radius."[1] In his preface to the 1958 edition of the published lectures, Allan Sandage referred to the evidence for the changes in the distance scale accumulated since Hubble delivered the lectures. "As of this writing [1958] the evidence suggests an increase of a factor of from 5 to 8 in Hubble's distance scale."

In terms of modern values for this scale, Hubble's estimates would imply that within the observational horizon of a telescope that could penetrate to the order of 4000 million light years, there would be some 100 million extragalactic systems. In a recent survey with one of the world's most modern telescopes,[2] capable of recording objects somewhat fainter

1. E. Hubble, *The Realm of the Nebulae* (New York, 1958), chap. 8.
2. The 48-inch Schmidt telescope at Siding Spring in Australia, photographing to a magnitude limit of +23. See R. J. Dodd, D. H. Morgan, K. Nandy, V. C. Reddish, and H. Seddon, in *Mon. Not. R. Astr. Soc.* 171 (1975): 329.

than Hubble's limit, this estimate of 100 million galaxies was quoted as the number of objects lying within the field of view. At this sensitivity limit the number of observable galaxies was still found to be increasing by two or three times per magnitude. We do not know for how long this increase in numbers will continue as observational techniques enable even fainter objects to be recorded. The largest modern telescopes, such as the 200-inch Hale telescope on Palomar and the Soviet 236-inch instrument in the northern Caucasus, when using contemporary electronic recording devices are believed to be capable of recording objects as faint as magnitude +25. Thus it seems that some 500 million to 1000 million extragalactic objects lie within the observational limits of our modern telescopes on Earth. When the large space telescope is placed in orbit in 1983 by the U.S. shuttle, the constraints imposed by the Earth's atmosphere will vanish and a further extension of sensitivity may be envisaged. Whether the increase in observable numbers of objects will be found to continue with this increasing sensitivity is a question of fundamental importance to cosmology.

Whenever the spectrum of one of these remote objects is determined, the identifiable spectral lines are found to exhibit an increase in wavelength. This redshift is interpreted by nearly all working astronomers as a cosmological effect arising from the expansion of the universe. Moreover, at the greatest distance of penetration there is still no unambiguous indication of any significant departure from the linear relation between distance and redshift first established by Hubble. This relation—the Hubble constant—is expressed as the ratio between the velocity of mutual recession of the galaxies and the distance between them. Hubble calculated the value of the constant as 530 km per sec per megaparsec but today the value of the constant is believed to lie between 50 and 100 km per sec per megaparsec (see chapter 11). If this cosmological effect implies that the universe is evolving from a more compacted state, as suggested also by the theoretical models based on the equations of general relativity, then it is possible to extrapolate backward in time to estimate how long ago the expansion began. This period of time, between the beginning of the expansion and the present, is known as the Hubble time.[3] If the expansion rate has been constant since the beginning, then the Hubble time is also the age of the universe. If the expansion rate was greater in the beginning, then the age of the universe is less than this Hubble time. Since no certain evidence has yet been found for this deceleration over cosmic time—that is, no agreed evidence exists for a departure from the linear relationship of the redshift/distance relationship—the Hubble time may be taken as indicative of the age of the universe.[4]

The original value for the Hubble constant of 530 km per sec per megaparsec implied a Hubble time of the order 2 billion years. This led to very serious difficulties in the interpretation of the observed universe within the framework of the evolutionary models of general relativity, because the age of the Earth, and of the Sun and stars in the Milky Way, was estimated to be much greater than 2 billion years. Subsequently the distance scales and the value of the Hubble constant were revised, and the present limits estimated for the Hubble time lie between about 10 billion and 20 billion years. Nevertheless, until 1952, when a major revision removed the apparent conflict between the age of the universe and the ages of the stars, an important stimulus was given to the development of a nonevolutionary cosmology in which the concept of an age for the universe lost significance.

The originators of this theory in 1948, Bondi and Gold,[5] and

---

3. It is also known as the Friedmann time.

4. Except in one class of solutions of the equations exhaustively developed and studied by Lemaître where the cosmological constant has a positive value. Lemaître considered that initially at time zero the "primeval atom" of the universe exploded. After a few thousand million years the force of this initial explosion was exhausted and the universe settled to a long period of stability filled with a uniform distribution of primeval gas. Eventually condensations occurred and the forces of cosmical repulsion (implied by the positive value of the cosmological constant) led to the expansion of the universe observed today. The Hubble time in this case relates to the beginning of this expansion and not to the long earlier period of the universe, which remains indeterminate either in theory or by observation.

5. H. Bondi and T. Gold, in *Mon. Not. R. Astr. Soc.* 108 (1948): 252.

Hoyle who gave it a mathematical formulation by a modification of the field equations of general relativity,[6] sought a
more logical solution to the cosmological problem than that
inherent in the conventional world models of general relativity. Their theory, which became known as the steady state, or
continuous creation, theory, evaded the problem of a singular
creation in the remote past by invoking the perfect cosmological principle—that on the large scale the universe exhibited a high degree of uniformity not only in space but also
in time. Because the expansion of the universe would ultimately move galaxies beyond the field of view of the observer, the
theory proposed that new matter was continuously created at
precisely the rate necessary to form new galaxies so that on the
large scale the universe always presented an unchanging appearance. Further, the logical argument implied that the theory brought the problem of origin or creation within the scope
of physical inquiry in contrast to the evolutionary theories
where the beginning emerged as a metaphysical concept forever beyond the scope of physical investigation or observation.

The theory of the steady state had these important attributes, even after the difficult question of the age of the universe had been removed from the evolutionary theories in
1952. The concept of continuous creation satisfied the condition imposed by the evolution of galaxies and the cosmological
expansion of the universe while accepting an infinite past and
infinite future for the unchanged large-scale structure of the
universe. For more than two decades after the theory was proposed, one of the most bitter conflicts in the history of
astronomy ensued between proponents of the steady state and
the evolutionary cosmologies. An important attraction of the
steady state cosmology was that it made specific predictions
about the condition of the universe that were open to observational test. For example, if one could observe the universe as it
existed in the remote past, then the predictions about the
number of galaxies per unit volume of space compared with

6. F. Hoyle, in *Mon. Not. R. Astr. Soc.* 108 (1948): 372.

the present value were markedly different. On the steady state theory the number should not vary at any epoch in the history of the universe. By contrast, evolutionary models imply a more densely packed universe at earlier epochs. With the development of radio astronomy and the improved techniques of optical observation, the necessary penetrations into the remote epochs of the universe became possible so that number counts of objects recorded should reveal these differences. However, uncertainties in observations were too great for a unified opinion to be formed, and both sides in the dispute continued to interpret the observations in terms of their own predilections.

It is doubtful whether any astronomer changed sides in this dispute until a completely new feature of the universe was revealed in 1965. At that time scientists at the Bell Telephone Laboratories and at Princeton, New Jersey, obtained observational evidence for an early state of the universe that seemed to indicate an initial condensed state.[7] A. A. Penzias and R. W. Wilson, when initiating the operation of a radio receiving system on a wavelength of 7 cm, primarily intended for space communication tests, discovered a background radiation isotropic over the sky to a few percent. Their interpretation of this as a relic radiation from the hot initial condition in the early state of the universe soon became widely accepted. They calculated the black body temperature to be 3.5 degrees absolute (later modified to 2.7 degrees). Apparently this radiation originated when the temperature of the universe was very high and had been propagating through space for almost the entire evolutionary phase of the universe (see chapter 14). The temperature observed today of 2.7 degrees is compatible with a redshift of the radiation appropriate to the cosmological expansion of the universe.

Although alternative suggestions have been made regarding

7. These results were published simultaneously by R. H. Dicke, P. J. E. Peebles, P. G. Roll, and D. T. Wilkinson of Princeton in *Astrophys. J.* 142 (1965): 414; and by A. A. Penzias and R. W. Wilson of Bell Telephone Laboratories in *Astrophys. J.* 142 (1965): 419. Penzias and Wilson had priority in the discovery and were awarded the Nobel Prize for physics in 1978.

the precise epoch in the early evolutionary phase when the radiation was emitted, no acceptable alternative has been proposed to the main thesis, namely, that this radiation is indicative of a very high temperature and highly condensed phase of the universe many billions of years ago. The discovery of this radiation therefore seems to provide conclusive evidence in favor of the generality of the world models of an evolutionary universe derived from relativistic theory, and it is decisively opposed to the concept of the steady state universe.

In the light of this evidence for the evolutionary universe, it is now assumed that the Hubble time is indicative of the time scale over which the universe has been expanding from the initial highly condensed state and that it represents the age of the universe if the expansion rate has remained constant over this cosmological time scale. The present estimates of the Hubble time lead to an age in the time span of 10 billion to 20 billion years. Are there other features of the universe that enable us to confirm this belief or place closer limits on the age?

One fruitful field of inquiry could possibly be that regarding the age of the local galaxy. We do not know the epoch of galaxy formation in the evolving universe, but it seems most probable that galaxies were formed in a limited period of time a few hundred million years after the beginning of the expansion. Thus we expect galaxies to be almost as old as the universe. The oldest objects in the local galaxy are the globular clusters; both observation and modern theories of star formation and evolution reveal that stars in globular clusters are much older than those in the solar neighborhood and in the spiral arms. Globular clusters contain a large proportion of stars that have evolved to the red giant stage, a process that calculations show will occur when some 10 to 15 percent of the hydrogen has been converted to helium. At this stage the stellar atmosphere will expand and cool, while the interior contracts and gets hotter until the thermonuclear transmutation of helium begins. By observation it is possible to estimate the mass of stars that have just reached this helium-burning stage in the globular clusters, and since the rate of consumption of the hydrogen is known, a calculation can be made of the age of these stars. This gives an age of 10 billion to 15 billion years for the globular clusters—and hence for the universe.[8]

Another important field of inquiry bearing on the age of the universe is the age of the chemical elements. The rate of disintegration of the naturally occurring radioactive elements is well established; for example, the half life of the uranium-238 isotope is 4.5 billion years and of the uranium-235 isotope 700 million years. All naturally occurring radioactive elements are daughter products of thorium, which has a half life of 13.9 billion years. If the abundance ratios are determined, these values provide a unique "radiogenic" clock from the time of the synthesis of the first elements. Since 1957, when Burbidge, Burbidge, Hoyle, and Fowler published the theory of the production of the elements in the interiors of stars by thermonuclear transmutation, there has been much debate on this issue.

When the local galaxy was formed from the primeval gas of the evolving universe, there were no heavy elements. When the solar system condensed from the interstellar medium some 4.5 billion years ago, however, this medium contained a quantity of heavy elements. These heavy elements had been produced in the interiors of the oldest stars and spread throughout the interstellar medium by various processes, particularly in supernova explosions. Thus, by determining the abundance ratios of the heavy elements in the Earth's crust, or in meteorites, it is possible to deduce the time at which they were formed. From the theory of supernova explosions of stars, a further extrapolation can be made to estimate how long ago the stars were formed from the primeval gas. Some of the most

8. This argument is based on the "standard model" for a star, which presumes that the stellar atmosphere is supported against collapse by internal gas pressure. The details of the calculations predict that nuclear reactions should produce neutrinos. Recently it was found that the neutrino flux from the Sun was at least three times less than that predicted by the standard model. Alternative models now under investigation, which could bring the predicted and observed neutrino flux into agreement, lead to changes in the amount of hydrogen that must be consumed before the red giant stage is reached, and hence would influence the age estimates derived in this manner.

recent work on this problem has been carried out by Fowler and his colleagues in California, who have developed a computer program to yield a self-consistent solution for the various complex events involved. A typical result of the calculations is that the galaxy was formed about 10.8 billion years ago and that over the next 6 billion years the region of space in which the solar system was to form accumulated 97.3 percent of its heavy elements. It is then assumed that a particular nearby event (e.g., a supernova explosion) contributed the remaining 2.7 percent of heavy elements.

One hundred and sixty million years after this, the solar system was formed and had evolved to such a state that the heavy elements had been locked in the meteorites we measure today. Taking this age for the solar system of 4.6 billion years, the age of the universe is obtained as the 10.8 billion years plus 1.5 billion years from the formation of the first stars. This gives an age for the universe of 12.3 billion years. In view of various uncertainties in the calculations, Fowler considers that the limits for the age lie between 10 billion and 15 billion years.[9]

It is a remarkable and rather beautiful feature of modern cosmology that the time scales for the age of the universe agree so well when derived by radically different procedures. Thus the age derived from the redshift measurements give a Hubble time lying between 10 billion and 20 billion years while that derived from considerations of galaxy formation and from the origin of the heavy elements both suggest an age of between 10 billion and 15 billion years. These time scales transport us to an epoch when there were no stars, no galaxies, and no heavy elements in the universe. Their close agreement suggests a logical connection between the cosmical expansion of the universe and the creation of stars and galaxies which, in combination with the evidence from the microwave background radiation, encourages a belief in the evolution of the universe from a singular condition in a remote epoch.

9. See for example W. A. Fowler, in *Bulletin of the American Academy of Arts and Sciences* 32 (November 1978): 32.

14

The Beginning and
End of the Universe

At the end of his Silliman lectures at Yale in 1935, Edwin
Hubble, who in the previous decade had obtained the decisive
evidence for the existence of extragalactic nebulae and for the
cosmological expansion of the universe, said this:

> Thus the explorations of space end on a note of uncertainty.
> And necessarily so. We are, by definition, in the very centre
> of the observable region. We know our immediate neigh-
> bourhood rather intimately. With increasing distance, our
> knowledge fades, and fades rapidly. Eventually we reach
> the dim boundary—the utmost limits of our telescopes.
> There, we measure shadows, and we search among ghostly
> errors of measurement for landmarks that are scarcely more
> substantial.[1]

Nearly a half century has gone since Hubble delivered those
lectures—years in which several technical developments have
revolutionized astronomy so that observers now have the ability to study the universe over the whole extent of the spectrum
instead of the limited octave of the visible region to which

1. *The Realm of the Nebulae* (New York: Dover, 1958), pp. 201–2.

Hubble was restricted. Theorists have powers of computation that in their capacity and speed had no parallel during Hubble's lifetime. Even so, Hubble's words remain a valid comment on the extent of our present knowledge of cosmology. The emergence of our understanding about the universe has been prolonged and tortuous with decisive turning points separated by long intervals of doubt and uncertainty.

Although many new discoveries have been made about the universe since the major large-scale structural features were revealed by Hubble's work, it is not easy to see that many of the major scientific problems of cosmology have been clarified. Some details have been clarified, but other discoveries have deepened the problem of the existence of the universe and our place in the evolution of the natural world. On the question of the age of the universe, for example, we discussed in chapter 13 the contemporary evidence that the present evolutionary phase began some 10 billion to 20 billion years ago. Although the large-scale harmony of these results is extraordinarily beautiful, the uncertainties about the relation of the Hubble time to the real age and to the epoch of formation of the galaxies are troublesome. The Hubble time is derived from the measurement of the redshift, which gives the rate of expansion of the universe. If this rate has been constant since the beginning of the expansion, then the derived Hubble time is simply related to the age of the universe. The uncertainties in this measurement give limits of between 10 billion and 20 billion years for the age. Unfortunately, it is not certain that the assumption of uniformity since the beginning of the expansion is justified.

A great deal of contemporary observational work on objects that lie at the limits of penetration of the telescopes is devoted to this issue. In the astronomical literature many differing interpretations are placed on the results of observations. Allan Sandage, one of the most renowned cosmologists of this era, in 1972 published apparently compelling evidence that the rate of the cosmological expansion had decreased since the earliest epochs. The figure he placed on this deceleration would imply that the age of the universe was very much less than that derived from the Hubble time. At the limits already given for the Hubble time of 10 billion to 20 billion years, the work of Sandage would imply that the age was about 40 percent less than these figures, between 7 billion and 11 billion years. Other interpretations also have been placed on the observations—in particular that redshift-distance measurements are uncertain because the galaxies used for the observations may have decreased in brightness with time. It is clear that we do not know precisely how the Hubble time is related to the age of the universe.

Regarding alternative approaches whereby the age is assessed from the calculation of the ages of galaxies or from the age of the heavy elements, both avenues indicate an age for the universe of between 10 billion and 15 billion years. Again this figure is sharply influenced by the assumption that the galaxies formed fairly early in the expansion of the universe. This assumption is based on the estimate that the expanding primeval gas *could* only condense into galaxies in a rather limited time span when the density of the universe was about ten thousand times its present value. This would mean that galaxies formed only a few hundred million years after the expansion began. It must be remarked that this is a purely theoretical estimate, and that observational evidence on this matter is effectively zero. In fact, observations of the most distant objects in the universe (the quasars) cast some doubt on the belief that galaxies condensed in the early stages of the expansion.

Many hundreds of quasars have been measured by the radio and optical telescopes since their discovery in the period from 1960 to 1963. In the most distant quasars yet detected the spectral lines have been redshifted so that their wavelength is increased by some five times. A rational interpretation of this is that the quasars (which appear to be associated with galaxies and probably occur early in the evolutionary phase of galaxy formation) did not form until the universe was about one fifth its present age. Thus the estimate of the age of the universe derived from the calculations of the age of galaxies and the heavy elements may need to be increased by 2 billion to 4 billion years instead of a few hundred million. Attempts are in progress to detect evidence of much earlier stages in the process of galaxy formation, but the results so far are negative.

This discussion has far wider implications in relation to the future history of the universe. Although the world models of general relativity seem to be in accordance with the universe of our observation, namely, that we exist in an evolving and expanding universe, undetermined constants in the equations make it impossible to specify the precise nature of the world model until the constants are determined by measurements of the actual universe. In particular, we cannot say from the present state of theory and observation whether the universe is open or closed. That is, whether the expansion will continue forever and the universe will eventually cease to exist except in a form of unavailable energy or whether the universe will eventually collapse to another superdense state similar to that from which it evolved. The physical alternatives are straightforward. If there is sufficient matter in the universe, then the expansion will eventually slow down because of the effects of gravitational attraction, and contraction will take place. If the density of matter in the universe is insufficient to slow down the expansion, then the universe is open and the expansion will continue forever (or reach a zero rate after infinite time). Although it is possible to calculate the critical density at the present epoch for various world models, we have no means of making a direct measurement of this density. Of course it is possible to calculate the density from counts of galaxies, the number of stars they contain, and the amount of interstellar gas and dust. This value for the density is insufficient to close the universe. A great deal of elegant work is in progress on this issue. Some cosmologists believe that the universe contains missing mass, at present unobservable, in sufficient quantity to raise the mean density above the critical value necessary to close the universe.

Although there is at present no answer to the question

about the future evolution of the universe, at least this remains a feature of cosmology open to observation. Only to a limited extent can this remark be made about its past history. With reservations about the uncertain effect of deceleration in the expansion of the universe, one can say that the observation of individual extragalactic objects to the most distant quasars provides a means of studying the universe over some three quarters of the elapsed time since the beginning of the expansion. On the present interpretation of the microwave background radiation being a relic radiation from the hot, dense state of the early universe, we look back over 99 percent of the time elapsed since the beginning of the expansion. Over the time interval, which may extend from hundreds of millions to a few billion years from that epoch, we have no observational data and little valid theory to guide us about the condition and events that led to the formation of galaxies as the universe evolved. Nevertheless, this remains an epoch open to future observation and study, at least in principle. The problem of immense difficulty in all its aspects arises when we inquire about the evolution of the universe in the earliest stages, that is, before the radiation was generated that we now observe as the microwave background.

The first question to ask abut this relic radiation is whether we can specify the conditions that existed in the early universe when it originated, and the time that had elapsed since the beginning of the expansion, when these conditions occurred. From the temperature of this microwave background radiation observed today, that is, 2.7 degrees absolute, it is a straightforward matter to calculate the density of the photons in space.

The answer is that there must be about five hundred photons per cubic centimeter of the universe. The important fact about this number is that it is vastly greater than the number of nuclear particles per cubic centimeter in the universe. We do not know precisely this number of particles because, as mentioned, we cannot measure the density. If the density is near the critical value necessary to close the uni-

verse, that is, for the gravitational attraction between the particles to overcome the forces of expansion, then the number of
nuclear particles is about one per million cubic centimeters of
space. The uncertainty about this number cannot obscure the
conclusion that throughout the universe the number of
photons that comprise the radiation, in a given volume of
space, is vastly greater than the number of nuclear particles
that comprise the matter of the universe. The discrepancy in
the number is about a billion times and may be even greater
if the matter density is less than that required to close the
universe.

Although this ratio of photons to particles is so great, the
energy balance in the present universe is dominated by the
particles in the universe. The equivalent energy in a particle
can be calculated from the Einstein relation ($E = mc^2$). Taking
the mass of a nuclear particle (a proton or neutron), this
energy is about 1000 billion times greater than the energy in
a photon at the temperature of the microwave background radiation. Thus the average energy balance in the universe today is at least a thousand times in favor of the particles. That
is, the universe is matter-dominated. If we accept the evidence
that the universe has evolved from a hot, dense state, this domination of matter over radiation could not have existed in the
early epochs of the universe. As the temperature increases, so
the energy per photon increases, although the particle
energies remain unaffected. A calculation indicates that the
photon energy would be the same as the particle energy when
the temperature of the universe was in the region of 3000 to
4000 degrees absolute, instead of the present observed value of
2.7 degrees absolute. At an earlier stage, when the universe
was at a higher temperature, the energy of photons exceeded
that of particles, i.e., the universe was radiation-dominated.

This transitional phase between the dominance of radiation
and the dominance of matter in the universe coincided with
the epoch when the energy of photons was no longer great
enough to prevent the permanent capture of electrons by nu-
clei to form atoms, that is, less than a few thousand degrees

absolute. At this stage the free electrons disappeared and the universe became transparent to radiation. We do not know why this transition from an opaque to a transparent condition occurred at about the same time as the transition from the domination of radiation to that of matter in the evolving universe. In any event, the interpretation favored today for the 2.7-degree radiation is that the photons belong to the earlier opaque phase of the universe, before this transitional phase occurred, and they have been propagated freely throughout the universe since that time. The temperature of this radiation has fallen in inverse proportion to the size of the expanding universe and is now 2.7 degrees. At the transitional phase, when the temperature was about 3000 degrees, the radius of the universe must have been a thousand times smaller than it is today and the density about a billion times greater.

In recent years several speculative accounts have been published in which the conditions in the earlier radiation-dominated phase of the universe have been described.[2] The conjectures are based on the assumption that the universe expanded uniformly in its earliest moments and that changes in density and temperature followed the well-established laws that govern the behavior of radiation today. The transition from the radiation-dominated to the matter-dominated phase is calculated to have taken place about 700,000 years after the beginning of the expansion. The main helium production in the universe is estimated to have occurred before the temperature had fallen to a billion degrees, 100 seconds after the expansion began. In the earliest minutes the temperature, calculated on this basis, would have been immensely higher so that the collision of the photons would have led to the creation of exotic fundamental particles. Soon after the formation of helium, when the temperature had dropped below a few billion

2. A concise popular account has been given by Steven Weinberg, *The First Three Minutes* (New York and London, 1977). The physical and mathematical background is the subject of many publications in scientific journals. For an authoritative work, see M. Rees, R. Ruffini, and O. A. Wheeler, *Black Holes, Gravitational Waves and Cosmology* (New York, 1974). A shorter work is J. Kleczek, *The Universe* (1976), esp. chap. 5.

degrees, the radiation could be considered to be expanding freely. The energy of the photons was then too low to lead to particle creation, and although photon-particle collisions continued, the excess of photons was such that after a few minutes, the era of radiation-domination began. The photons now measured in the 2.7-degree background radiation are believed to be relic photons from this opaque, radiation-dominated phase when the temperature was decreasing from a few billion degrees as the universe expanded.

On this interpretation of the origin of the relic radiation photons a remarkable feature of the early universe is revealed by the measurements. The study of the distribution of the strength of this radiation should reveal any departure from uniformity that existed in the universe during the era of radiation domination, that is, from a few minutes to 700,000 years after the beginning of the expansion, at which time the matter and radiation became effectively decoupled. The contemporary measurements of the radiation fail to reveal any departure from a uniform isotropic distribution over the heavens, and so we are led to conclude that in the very early radiation-dominated phase the universe was isotropic at least to one tenth of one percent. We do not know the extent to which our conventional time scales have meaning in those remote epochs. Indeed, although we are forced to use conventional language, the relation of these early events to the scale of minutes and years established billions of years later, when the solar system was formed, is without meaning. Even so, this uncertainty does not affect the compelling evidence that the relic radiation measured today implies that the universe evolved from a hot and densely compacted radiation phase some 10 billion to 20 billion years ago and, furthermore, that there was a high degree of uniformity throughout the universe in the earliest stages of the expansion.

This comment about the isotropy of the early universe has a critical bearing on the consideration of the initial state—the beginning of time and space. The world models derived from the general theory of relativity provide a mathematical formulation of the change in the radius of curvature of the universe with time. The various solutions all have the common feature that they predict a singular condition for the universe at time zero: a beginning of zero radius and infinite density. It has been possible to escape from this physical dilemma by postulating that the singularity is a mathematical difficulty introduced by the assumption of uniformity in the universe. The measurements on the early universe now seem to close this avenue of escape from the difficulty of providing a physical description of the universe at time zero, and it is necessary to inquire to what extent theories of modern physics enable an extrapolation to be made toward the zero of time, even as a matter of principle.

This answer is provided by the limits of extrapolation of the quantum theory, which defines a "characteristic length" involving the fundamental parameters of the gravitational constant, the velocity of light, and Planck's constant. This length is 2.6×10^{-33} cm. The mathematics of the world models imply that the expanding universe would have reached this size 10^{-43} seconds after the beginning of the expansion. The application of the known physical laws then imply that the density of the universe would have been 5×10^{93} gms per cc and the temperature 10^{33} degrees absolute. Although these figures are impossible to comprehend, they have this great significance: they represent the limits to which we can trace the earliest moments of the universe within the scope of physical theory.

We have no physical language with which to describe the state of matter when quantum fluctuations of the gravitational field become significant in this manner. This problem of quantizing the gravitational field has defied decades of intellectual effort. There are differing reactions to this failure. Some feel that a solution will eventually be discovered within the framework of existing fundamental theories. Others take the view that the theories of general relativity and quantum physics are incompatible. In this case either the present theories of space-time need fundamental revision or a physical description of the earliest state of the universe is forever impossible.

15

The Harmony of the World

The belief that the world must perforce exhibit a fundamental harmony has been a powerful influence on the emergence of cosmology. The strength of belief in the perfection of the circle and the sphere has been remarkable, to the extent that cosmology was dominated by this tenacious attitude until Kepler derived his laws of planetary motion early in the seventeenth century. In the *Timaeus* Plato bases his cosmology on the belief that the world is a globe because like is fairer than unlike, and only a globe has the property of being alike everywhere.

> Wherefore He made the world in the form of a globe, round as from a lathe, having its extremes in every direction equidistant from the centre, the most perfect and the most like itself of all figures: for He considered that the like is infinitely fairer than the unlike.

> Moreover, this globe rotates because circular motion is the most perfect motion. The four elements—fire, air, water, and earth—were harmonized by God in the correct proportions when He made the world. Therefore the world is perfect and indissoluble except by God. Although the cosmology of his pupil, Aristotle, differed in important respects (e.g., Aristotle,

unlike Plato, did not believe that time was created), the priority of the circle was retained. Aristotle held that circular motion was of the primary kind, and only circular motion could be continuous and infinite. The Earth was a sphere at the center of the universe, and the homocentric set of nesting shells was filled with an ether whose circular rotations accounted for the celestial and planetary motions.

This fervent belief in the perfection of the circle and the sphere led to the evolution of the complex Ptolemaic epicyclic universe. Furthermore, the underlying motive in the revolutionary scheme of Copernicus was the desire to remove the artificial device of the equant. As late as the sixteenth century, the difficult question of the motion of the Earth was less disruptive than the abandonment of the perfection of the circle and the uniformity of motion of the heavenly bodies. In the prefatory letter to Pope Paul III with which *De revolutionibus* opens, Copernicus emphasized that the heavens "became so bound together that nothing in any part thereof could be moved from its place without producing confusion of all the other parts of the Universe as a whole. . . ." The Copernican system possessed a beautiful harmony, not merely in the preservation of circular motions but also because the relative dimensions of the system were defined in the Sun-centered universe. But for this aesthetic beauty the heliocentric theory may not have survived at that epoch, for the predictions of the Ptolemaic theory were at least as accurate, and the evidences of the senses as well as the Scriptures were utterly opposed to the new cosmology.

This fervent belief in the harmony of the world dominated the outlook of Kepler. Nowhere is this almost mystical feeling for the orderliness of nature more evident than in the tortuous paths that led him to his three laws of planetary motion. In his first book, the *Mysterium cosmographicum* of 1596, he describes his vision that the universe must be built around the five regular solids. In retrospect it is easy to dismiss this as irrelevant, but the model illuminated for Kepler an important feature: the orbital planes of the planets intersected in the Sun and not in the Earth as in the Copernican model. His long search for harmony becomes even more explicit twenty-three years later with the publication in 1619 of *Harmonices mundi* in which he enunciates the third law relating the period of a planet to its average distance from the Sun. *Harmonices mundi* is a famous work because it contains this third law of planetary motion, but the book is not a treatise specifically devoted to the derivation of this law. In fact, it is a discussion about proportions, intervals, and harmonic procedures. Kepler shows, for example, that the angles the line joining a planet to the Sun describes at the aphelion and perihelion passages form a complete system of concordant musical intervals. There are thirteen theorems of this type in the work. Twelve are either erroneous or irrelevant to the planetary system as subsequently understood, but the eighth is the historic third law of planetary motion.

For two millennia the perfection of the circle, circular motion, and the sphere were dominating criteria, leading first to the invention of epicycles and deferents and eventually, with Copernicus, to the abandonment of the geocentric concept of the stationary Earth. Suddenly, with Kepler, another feature becomes dominant. The harmony of the numerical relations assumes a greater importance than the perfection of the circle. This is one of the unique features in the emergence of cosmology, when the relationship of numbers assumes a more sublime importance than geometric features. Kepler wrote:

> When there is a choice between different things which are not completely compatible with each other, preference must be given to the one which has higher status, and the lower one must be abandoned in so far as this is necessary. This is evident from the very word Cosmos, which means the beauty of order. In the same way that life ranks higher than the body, and Form higher than matter, harmonic beauty takes precedence over the beauty of simple geometry.[1]

1. This quotation from Kepler's *Weltharmonik*, trans. Max Caspar, has been taken from the article by Rudolf Hasse in *Kepler–Four Hundred Years*, ed. A. and P. Beer (New York, 1975) p. 527.

The mathematical formulation of the Keplerian laws on the basis of the concept of universal gravitation by Newton emphasized the large-scale harmony of the world. But there were underlying problems. Newton did not calculate the mutual attraction between the planets; he estimated the effects on the basis of the perturbations caused by the Sun in the orbit of the Moon around the Earth. This led him to overestimate the effect of these perturbations, and in order to preserve the harmony and long-term stability of the planetary system he had recourse to divine intervention. Thus in the *Opticks* he wrote:

> . . . the hard and solid Particles above mention'd, variously associated in the first Creation by the Counsel of an intelligent Agent. For it became him who created them to set them in order. And if he did so, it's unphilosophical to seek for any other Origin of the World, or to pretend that it might arise out of Chaos by the mere Laws of Nature; though being once form'd, it may continue by those Laws for many Ages. For while Comets move in very excentrick Orbs in all manner of Positions, blind Fate could never make all the Planets move one and the same way in Orbs concentrik, some inconsiderable Irregularities excepted which may have risen from the mutual Action of Comets and Planets upon one another, and which will be apt to increase, till this System wants a Reformation. Such a wonderful uniformity in the Planetary System must be allowed the effect of choice.[2]

Indeed, for Newton the beautiful harmony of the world emphasized the divine nature of Creation, as the perfection of the circular motion did for the Greek philosophers. In the *General Scholium* to book 3 of the *Principia* he wrote:

> The most beautiful system of the Sun, planets, and comets, could only proceed from the counsel and dominion of an

---

2. *Opticks* (2nd ed.; London, 1717), query 31. This passage does not occur in the 1st ed. (1704).

intelligent and powerful Being. And if the fixed stars are
centres of other like systems, these, being form'd by the like
wise counsel, must be all subject to the dominion of One;
especially since the light of the fixed stars is of the same
nature with the light of the Sun, and from every system light
passes into all other systems: and lest the system of the fixed
stars should, by their gravity, fall on each other, he hath
placed those systems at immense distances from one
another.[3]

More than two centuries elapsed before the underlying harmony of the Newtonian world was questioned. On November
6, 1919, in the rooms of the Royal Society in London,[4] Sir
Frank Dyson, the Astronomer Royal, announced the results of
the expedition to Sobral in West Brazil and to the island of
Principe off the west coast of Africa. Expedition members,
during the total solar eclipse of May 29, 1919, had attempted
to verify the prediction of the general theory of relativity regarding the amount of the deflection of a ray of light from a
star as it grazed the solar disk. Einstein had made three predictions by which the general theory could be tested observationally. One, concerning the advance of the perihelion of
Mercury, had been found to be correct. The test of a second
prediction, that the spectral lines from a dense star should be
shifted in wavelength because of the intense gravitational field,
had at that time given inconclusive results.[5] Understandably,
because of the revolutionary concepts and inferences of the
general theory, there was deep interest when Dyson announced that the measurements had revealed a deflection of

3. *Principia*, bk 3, *The System of the World* p. 544 (1947 Cajori revision of Motte's 1729
translaton). The reference in this extract to "immense distances" was later clarified
in the Newton-Bentley correspondence to imply the infinite extent of the universe (see
chap. 9). The differing references to God in the successive editions of *Principia* have
been discussed by Cohen. See I. B. Cohen, *Isaac Newton's Principia, the Scripture and
Divine Providence*, in *Essays in Honor of Ernest Nagel: Philosophy, Science and Method* (New
York, 1969).
4. A joint meeting of the Royal Society and the Royal Astronomical Society. The
President of the Royal Society, Sir J. J. Thomson, was chairman of this meeting. At
that time, the rooms of the Royal Society were in Burlington House.
5. The prediction was eventually verified in 1925 by W. S. Adams of Mt. Wilson in
the shift of the spectral lines from the white dwarf star Sirius B.

starlight twice as great as that to be expected on Newtonian theory, and therefore in accordance with Einstein's predictions. A. N. Whitehead, who was present at this meeting, compared the atmosphere with that of a Greek tragedy. "The absorbing interest in the particular heroic incidents, as an example and a verification of the workings of fate, reappears in our epoch as concentration of interest on the crucial experiments . . . in the background the picture of Newton to remind us that the greatest of scientific generalisations was now, after more than two centuries, to receive its first modification. . . ."[6]

Indeed, the particulate harmonies of the Newtonian world had been replaced by a universal harmony in which the geometry of space was linked to the material of the universe. The morning after the eclipse results were made public, the lead writer in *The Times* wrote:

From Euclid to Kepler, from Kepler to Sir Isaac Newton, we have been led to believe in the fixity of certain fundamental laws of the Universe. The centre of a circle was always equidistant from all points of its circumference. The sum of the angles of every triangle was always two right angles. On such belief practice and philosophy were based . . . the ordered arrangement of suns and planets in their courses were based on it . . . enough has been done to overthrow the certainty of ages and to require a new philosophy of the Universe, a philosophy that will sweep away nearly all that has hitherto been accepted as the axiomatic basis of physical thought.[7]

About this cosmic harmony of the universe he had revealed, Einstein wrote of the scientist as one for whom "religious feeling takes the form of a rapturous amazement at the harmony of natural law, which reveals an intelligence of such superiority that compared with it, all the systematic thinking and act-

---

6. A. N. Whitehead, *Science and the Modern World* (London, 1926), chap. 1.
7. "The Fabric of the Universe," The Times (London) Nov. 7, 1919, p. 13.

ing of human beings is an utterly insignificant reflection.''[8]

Telescopes of the modern world have revealed to us a universe that, at least on the large scale, seems to be compatible with the world models of general relativity. Although this beautiful harmony exists in the cross section of the world we observe today, many severe problems remain. The beginning and the end seem physically indeterminate and incomprehensible both in terms of theory and observations. A most surprising feature of the evolving universe, defined by modern theory and observation, is the evidence for its homogeneity and isotropy in the earliest stages of the expansion. Although the world models predict a nonstatic universe unless it is devoid of matter, no features underlying the theories impose limitations on the rate of this expansion—and this rate is critical for the emergence of stars, galaxies, and man from the primeval state. Underlying all is the difficulty with the singularity at zero time in the relativistic models, and the consequent failure to construct a theory of space-time on the basis of relativity and quantum mechanics applicable to the state of the universe at the earliest moments of the expansion.

Notwithstanding the great success of the general theory, it is for these reasons that theorists search for an even deeper understanding of the natural world. The relation of the Einstein theory to Mach's principle has been questioned. Einstein believed that the field equations of general relativity incorporated Mach's principle without ambiguity, but alternative solutions of the equations have been discovered. Modifications to Einstein's equations have been proposed that do give exact expression to Mach's principle—they differ in certain predictions such as the amount of deflection of starlight grazing the solar disk. In these respects very precise measurements in recent years have been in support of the Einstein theory. The topological structure of space-time, and the question of the involvement of Mach's principle as a determinant

8. Albert Einstein, *The World as I See It* (London, 1935). Translated from the German *Mein Weltbild*.

in the nature of space, are complex issues facing contemporary theorists.

A number of cosmologists engaged in a search for the harmony relating the microphysical to the macrophysical world —that is, the scale of the atom to that of the universe—have drawn attention to the existence of a series of strange numerical relationships. In this modern version of the search for the harmonies of the world the question is not that of extraneous relationships between concordant musical intervals and the structure of the planetary system, which inspired Kepler. On the contrary, it has been demonstrated that the dimensions of the world, from the microstructure of the atom to the scale of the universe, are apparently determined by a small number of well-known physical constants.

Eddington was a distinguished pioneer in the search for the numerical relationships he believed could unify the physical laws governing atomic structure with the large-scale structure of the universe. In major works published in 1936 and 1946[9] Eddington attempted to demonstrate that the numbers 136 and 137 have particular importance in the space of the Einstein universe and emphasized the close relationship of these numbers to a dimensionless number in atomic physics involving only the velocity of light, the charge on the electron, and Planck's constant.

At the time, Eddington and Lemaître had developed one of the world models of general relativity involving a positive value for the cosmological constant. On the assumptions of this world model, Eddington showed that as time in the expanding universe approached infinity, the theoretical value for the Hubble constant would be the square root of one third of the cosmological constant. Eddington also derived a numerical value for this quantity from the fundamental atomic constants. He calculated this to be 572.4 km per sec per mega-

9. A. S. Eddington, *Relativity Theory of Protons and Electrons* (Cambridge, 1936); idem, *Fundamental Theory* (Cambridge, 1946). Eddington made an attempt to explain these complex ideas in a more popular work, *The Expanding Universe* (Cambridge, 1933), chap. 4. Eddington, born in 1882, died in 1944. *Fundamental Theory* was published posthumously.

parsec, a result remarkably close to the value of 540 km per sec per megaparsec that Hubble had derived from observations of the universe. Further, in this world model, space was finite. Eddington calculated that the number of particles in the universe was simply related to the radius and to the number 136, and that this number was 2.4×10^{79}. It is remarkable that if certain fundamental parameters of the world are taken—specifically the charge and mass of the electron, the mass of the proton, the gravitational constant, the velocity of light, the critical density of the universe, and the Hubble time—they can be combined to give a set of different pure numbers. One of these, which expresses the ratio of the electrical to gravitational force between an electron and a proton, is 0.23×10^{40}, the square of which (5.3×10^{78}) is within a factor of 4 of Eddington's figure for the number of particles in the universe. Two further pure numbers involving the Hubble time and the atomic constants are 4×10^{40} and 10^{80}, again remarkably close to the square root and the number of particles in the universe as calculated by Eddington.

It is scarcely credible that the close relationships between these numbers derived from the constants of atomic structure and the scale of the entire universe can be accidental. Indeed the chance of coincidences between numbers of the order of 10^{40} derived in this manner can be disregarded. It is difficult to resist the conclusion that they represent some deep relation in the universe. Eddington certainly believed this to be the case: ". . . nature's curious choice of certain numbers such as 137 in her scheme—these and many other scraps have come together and formed a vision."[10] Eddington's mathematical derivation of the numerical relationships was extremely involved, and very few scientists have been able to comprehend his *Fundamental Theory* to the extent necessary to pursue his vision into the realms of modern cosmology. In particular, the special world model of the Eddington-Lemaître universe, with the positive cosmological constant, is no longer believed to repre-

10. *The Expanding Universe*, p. 126.

sent the universe as observed. Nevertheless, the strange and close relationships of the pure numbers derived from the microphysical and macrophysical world remain of compelling interest.

At the time when Eddington was concerned with the meaning of these large numbers, P. A. M. Dirac developed a theoretical cosmology in which he proposed that large numbers of the order 10^{40} deriving from the fundamental constants had varied with the age of the universe.[11] In this theory the age of the universe is expressed in units of time determined by the electronic charge and mass, Planck's constant, and the velocity of light. The present number derived is of the order 10^{40}, which is also the ratio of the electrical to gravitational forces between an electron and proton. The theory implies that this figure changes with the age of the universe, and that the gravitational constant varies inversely as the age. The general theory of relativity assumes the stability of the gravitational constant throughout all time, and in order to reconcile the Dirac cosmology with general relativity it must be assumed that the number of particles in the universe increases with time.

The theory makes a number of revolutionary predictions. For example, since the wavelength of the light observed from a distant galaxy is determined by the atomic constants at the time it was emitted, the redshift can be explained without invoking the expansion of the universe. This prediction would imply that the universe is static; hence, if the theory is found to be correct, we would experience one of the greatest cosmological revolutions in history. A critical test of the theory concerns the evidence for the stability of the gravitational constant. The predicted rate of change is only a few parts in 100 billion per year. So far the most accurate measurements have set limits about ten thousand times more than this value. It is possible that future developments will enable closer limits to be placed on the stability of the gravitational constant, al-

11. P. A. M. Dirac, in *Nature* 139 (1937): 323. See also *Proc. Roy. Soc. A.* 165 (1938): 199.

though it must be remarked that few astronomers today view Dirac's theory with favor.

The most recent developments in the search for the underlying harmonies of the world embrace the presence of ourselves as observers in the universe. The constants of nature, whether microphysical or cosmic, are those *we* measure. The proposition that what we can expect to observe must be restricted by the conditions necessary for our existence and our presence in the universe as observers has been termed the *anthropic principle*.[12] The evolution of ourselves, or of other sentient beings in the universe, depends on many features in the large-scale evolution of the universe that were narrowly determined by conditions in the universe in the earliest stages of its evolution. For example, the proportions of helium and hydrogen were determined in the first minutes of the expansion. If the forces of attraction between the fundamental particles had been slightly stronger, helium would have been dominant. In this event no stars of the type we observe with sufficiently long-term stability to facilitate evolution would have existed.

A similar argument applies to the rate of expansion of the universe. If this had been marginally slower or faster, the hydrogen that remained when matter and radiation became decoupled would not have formed into stars and galaxies. Either the universe would have collapsed before galaxy formation occurred or the primeval material would have dispersed. The time scale we measure for the evolution of the universe, and for the formation of galaxies and stars, closely determines the possibility of organic evolution. The existence of gravitationally bound systems such as stars and galaxies are essential to the evolution of life. The constants that determine these large-scale features of the universe show these extraordinary large-number coincidences with the fundamental

12. One of the first uses of this term seems to have been in the contribution of B. Carter to the IAU Symposium held in Poland in 1973 in honor of the quincentenary of the birth of Copernicus. See B. Carter, "Large Number Coincidences and the Anthropic Principle in Cosmology," in *Confrontation of Cosmological Theories with Observational Data* (Reidel, 1974), p. 291.

atomic constants on which, ultimately, the elements necessary for our existence were produced.

A more detailed and extensive investigation[13] of the relationships leads to the conclusion that the universe that exists, *and* our presence in the universe as observers, is critically contingent on the value of only a few physical constants and that these constants on the atomic and cosmical scales exhibit unique large-number relationships. Inevitably this leads to the question whether a multiplicity of universes exists, in which the physical constants differ from those of our world and hence cannot be known because they do not have the characteristics that appear to be essential for organic evolution.

When theory and observation were first revealing the nature of the evolving universe, A. N. Whitehead in his Lowell lectures of 1925 foresaw this present state of cosmology:

"There is no parting from your own shadow. To experience this faith is to know that in being ourselves we are more than ourselves: to know that our experience, dim and fragmentary as it is, yet sounds the utmost depths of reality: to know that detached details merely in order to be themselves demand that they should find themselves in a system of things: to know that this system includes the harmony of logical rationality, and the harmony of aesthetic achievement: to know that, while the harmony of logic lies upon the universe as an iron necessity, the aesthetic harmony stands before it as a living ideal moulding the general flux in its broken progress towards finer, subtler issues."[14]

---

13. A comprehensive review of the extraordinary relationships manifested throughout the entire scale of the cosmos from atoms through cells, man, planets, stars, and the universe has recently been given by B. J. Carr and M. J. Rees, "The Anthropic Principle and the Structure of the Physical World," *Nature* 278 (1979) 605.
14. *Science and the Modern World* (London, 1926), pp. 23–24.

Index

About The Author

Sir Bernard Lovell, O.B.E., LL.D., D.Sc., and Fellow of the Royal Society of Great Britain, is professor of radio astronomy at the University of Manchester and director of the Experimental Station at the Nuffield Radio Astronomy Laboratories, Jodrell Bank, Macclesfield, Cheshire, England. He is the distinguished author of several books and many articles dealing with space exploration, the nature of the universe, the origins of life and man's relationship to the cosmos. He also lectures widely on the implications of scientific growth and its consequences for man and the universe, showing with compelling clarity and originality the convergence of man with nature.

About The Founder of This Series

Ruth Nanda Anshen, Ph.D., philosopher, author, and editor, founded, plans, and edits several distinguished series, including World Perspectives, Religious Perspectives, Credo Perspectives, Perspectives in Humanism, The Science of Culture Series, The Tree of Life Series, and Convergence. She also writes and lectures on the relationship of knowledge to the nature and meaning of man and to his understanding of and place in the universe. Dr. Anshen's book, *The Reality of the Devil: Evil in Man,* a study in the phenomenology of evil, is published by Harper and Row. Dr. Anshen is a member of the American Philosophical Association, the History of Science Society, the International Philosophical Society and the Metaphysical Society of America.